HOW TO FACE 'THE FACES' OF CARDIAC PACING

How to face 'the faces' of CARDIAC PACING

edited by

ERIK ANDRIES
Department of Cardiology, O.L.V. Hospital, Aalst, Belgium

PEDRO BRUGADA
Department of Cardiology, O.L.V. Hospital, Aalst, Belgium

and

ROLAND STROOBANDT
Department of Cardiology, St Jozef Hospital, Oostende, Belgium

Lederle cardio
CYANAMID

Springer-Science+Business Media, B.V.

First published 1992
Reprinted 1993

ISBN 978-94-010-5139-2 ISBN 978-94-011-2582-6 (eBook)
DOI 10.1007/978-94-011-2582-6

Printed on acid-free paper

Contents

Preface

How to face 'the faces' of cardiac pacing represents an editor's compiled selection of lectures on cardiac pacing and electrophysiology.

Electrical stimulation of the heart is an ever-changing and, at times, explosive field. The number of implanting centres is growing tremendously and pacing is not exclusively confined to arrhythmologists. Therefore, the editors attempted to organize a course being both practical in daily clinical management and instructive in understanding technical concepts.

The glossary of terms have to be clearly understood before one is able to interpret the complex electrocardiograms of DDD and especially DDDR pacemakers. Those electrocardiograms have to be approached in a systematic way, using a step-by-step analysis.

The main clinical symptom requiring pacemaker implantation is syncope. It cannot be over-emphasized that syncope is a clinical diagnosis merely based on history and physical examination.

The organization of a pacemaker follow-up clinic depends on local facilities and needs. The effectiveness of pacing controls markedly increases when using a systematic approach. Repeated optimal adjustment of programmable functions is part of the control.

Antiarrhythmic drugs are loosing popularity in the treatment of tachyarrhythmias. Nonpharmacologic treatment (antitachypacing, implantable defibrillators and antiarrhythmic surgery) at the present time have definite indications, probably expanding in the future.

When complexity in electronic devices increases, repercussions on expenses, either by the government or social and private insurances, needs consideration.

We are grateful to the contributors of this volume. By editing this text, we have learned a great deal, and we hope you will enjoy facing the 'faces' of pacing.

September 1991

Erik Andries,
Pedro Brugada,
Roland Stroobandt

E. Andries, P. Brugada & R. Stroobandt (eds.), How to face 'the faces' of cardiac pacing, vii.

List of contributors

Erik ANDRIES
Department of Cardiology, O.-L.-Vrouwziekenhuis, Moorselbaan 164, B-9300 Aalst, Belgium

Co-authors: Hilde Willems, Sinan Gürsoy, Paul Nellens, Marc Goethals and Pedro Brugada

Pedro BRUGADA
Cardiology Centre, O.-L.-Vrouwziekenhuis, Moorselbaan 164, B-9300 Aalst, Belgium

Co-authors *Chapter 2*: Sinan Gürsoy, Jacob Atié, Marc Goethals, Günter Steurer, Hilde Willems, Josep Brugada and Erik Andries

Co-authors *Chapter 14*: Francis Wellens, Paul Nellens, Sinan Gürsoy, Jacob Atié, Günter Steurer, Erik Andries and Hugo van Ermen

A. JOHN CAMM
Department of Cardiological Sciences, St. George's Hospital, Medical School, Cranmer Terrace, London SW17 0RE, U.K.

Co-author: Clifford J. Garratt

Luc DE ROY
Department of Cardiovascular Pathology, University Clinic UCL de Mont-Godinne, Avenue du Dr Therasse, 1, B-5530 Yvoir, Belgium

Marc GOETHALS
Cardiology Centre, O.-L.-Vrouwziekenhuis, Moorselbaan 164, B-9300 Aalst, Belgium

Co-authors: Willy Timmermans, Roger Willems, Erik Andries and Roland Stroobandt

Hein HEIDBÜCHEL
Laboratory for Electrophysiology, University of Louvain, Gasthuisberg, Herestraat 49, B-3000 Louvain, Belgium

Co-author: Hugo Ector

Luc JORDAENS
Department of Cardiology, University Hospital Ghent, De Pintelaan 185, B-9000 Ghent, Belgium

Co-authors: Patrick Vertongen, Etienne Van Wassenhove and Denis L. Clement

Jean François LECLERCQ
Department of Cardiology, Lariboisière University Hospital, 2, Rue Ambroise Paré, F-75010 Paris, France

Co-authors: Isabelle Denjoy, Antoine Leenhardt and Philippe Coumel

Morton M. MOWER
Cardiac Pacemakers Inc., 4100 Hamline Avenue North, St. Paul, MN 55112-5798, U.S.A.

Co-author: Seah Nisam

Alfons SINNAEVE
Technical University of Ostend, Zeedijk 101, B-8400 Oostende, Belgium

Roland STROOBANDT
Dept. of Cardiology, St. Jozef Hospital, Nieuwpoortsesteenweg 57, B-8400 Oostende, Belgium

Co-authors *Chapters 7 and 13*: Roger Willems and Alfons Sinnaeve

Rob Th. M. VAN DEN OEVER
Department Insurance Medicine, School of Public Health, University of Louvain, Kapucijnenvoer 35, B-3000 Louvain, Belgium

Andre WALEFFE
Department of Cardiology, University Hospital, Sart-Tilman, B-4001 Liège, Belgium

1. A synoptic view on syncope

ERIK ANDRIES, HILDE WILLEMS, SINAN GÜRSOY,
PAUL NELLENS, MARC GOETHALS & PEDRO BRUGADA

INTRODUCTION

Definition. Syncope is defined as a transient loss of consciousness with spontaneous recovery. Syncope should not be confused with dizziness, unsteadiness, coma or shock. Epileptiform seizures usually are absent. However, they do occur when the attack is prolonged. Loss of consciousness in the presence of focal neurologic signs should be defined as a transient ischemic attack (TIA).

Incidence. Syncope is a common medical problem occurring in approximately 20% of the population [1, 2]. Although it can be assumed that most people with a syncopal episode don't seek medical attention, nevertheless, syncope accounts for 3% of the emergency room admissions and 1% of the total hospital admissions in the USA. When recurrent, it may account for more than 10% of the total population undergoing invasive electrophysiological investigations.

CAUSES OF SYNCOPE

Syncope has many causes, ranging from benign disorders (e.g. vasodepressor or hyperventilation syncope) which are usually clinically unimportant, to serious conditions (e.g. complete heart block, ventricular tachycardia) which, if unrecognized, may result in fatality. Today's syncope may represent tomorrow's sudden death.

Syncopal attacks are transient, so patients are usually seen when recovered and the cause often is not apparent. There is a high spontaneous resolution rate in up to two thirds of untreated patients with syncope. In the same patient, the repetitions of syncope may have more than one cause. One should be aware that simply finding an obvious cause (e.g. postural hypotension secondary to overmedication) should not stop the search for other

1

E. Andries, P. Brugada & R. Stroobandt (eds.), How to face 'the faces' of cardiac pacing, 1–21.
© 1992 *Kluwer Academic Publishers, Dordrecht*

serious and potentially life-threatening causes that can be treated. Some of the most common reported causes of syncope are summarized in Table 1.

Despite extensive evaluation, the cause of syncope remains unexplained in 38 to 47% of the patients studied. Obviously, estimates of the incidence of syncope of unknown origin are dependent on the thoroughness of previous investigations, and on the selection of patients.

Cardiovascular causes

1. *Obstructive cardiac lesions*

Obstruction to blood flow is able to occur anywhere in the heart (left ventricular inflow and outflow, right ventricular inflow and outflow), but also in the pulmonary circulation (pulmonary emboli, pulmonary hypertension) (Table 1).

The majority of chronic obstructive cardiac lessions cause syncope during effort. In pedunculated myxomata, syncope might occur during changes in body posture.

Aortic valvular stenosis is the most common obstructive lesion causing

Table 1. Common causes of syncope.

Cardiovascular

1. *Obstructive cardiac lesions*
   - Left ventricular outflow
     - — Aortic stenosis
     - — Hypertrophic cardiomyopathy
   - Left ventricular inflow
     - — Mitral stenosis
     - — Myxoma
     - — Thrombus
   - Right ventricular outflow
     - — Pulmonary valve and subpulmonic stenosis
   - Right ventricular inflow
     - — Tricuspid stenosis
     - — Myxoma
     - — Thrombus
   - Pulmonary circulation
     - — Emboli
     - — Pulmonary hypertension
     - — Pulmonary artery stenosis
   - Prosthetic valve malfunction
2. *Postural hypertension*
   - Often secondary to overmedication, dehydratation or acute blood loss.
3. *Vagally mediated syncope*
   - Vasovagal attack (common faint)
   - Carotid-sinus hypersensitivity
   - Post-tussive syncope
   - Micturition syncope
4. *Arrhythmias*
   - Sick sinus syndrome
   - High-grade atrioventricular block
   - Ventricular tachycardia
   - Ventricular fibrillation
   - Torsades de Pointes
   - Supraventricular tachycardias
   - Atrial fibrillation and flutter

Neurologic
- Seizure disorder
- Transient ischemic attack
- Subarachnoid hemorrhage

Metabolic and Endocrine
- Hypoglycemia
- Hypoxemia
- Addison's disease
- Pheochromocytoma

syncope. The adult patient with acquired aortic valvular stenosis has an excellent prognosis when asymptomatic (Figure 1). However, once symptoms appear, survivorship is markedly reduced in unoperated patients [3]. One can expect approximately 50% of patients who develop syncope to be dead within 5 years unless valve replacement is performed. Therefore, in patients with acquired aortic valvular stenosis, the occurrence of syncope is an alarming sign and valve replacement should not be delayed.

In hypertrophic cardiomyopathy, both mechanical and electrical disturbances can cause syncope. Their relative role is unclear at the present time. The benefit of antiarrhythmic drug therapy in asymptomatic patients with frequent VPB's or nonsustained ventricular tachycardia is controversial. Antiarrhythmic drugs can impair hemodynamics [4], or cause atrioventricular conduction disturbances. On the other hand, some antiarrhythmic drugs, such as calcium channel blockers, may have beneficial hemodynamic effects in this condition.

2. *Postural hypotension*

Postural hypotension mainly occurs during drug intoxication, dehydratation or acute blood loss. A wide variety of autonomic neuropathies, both

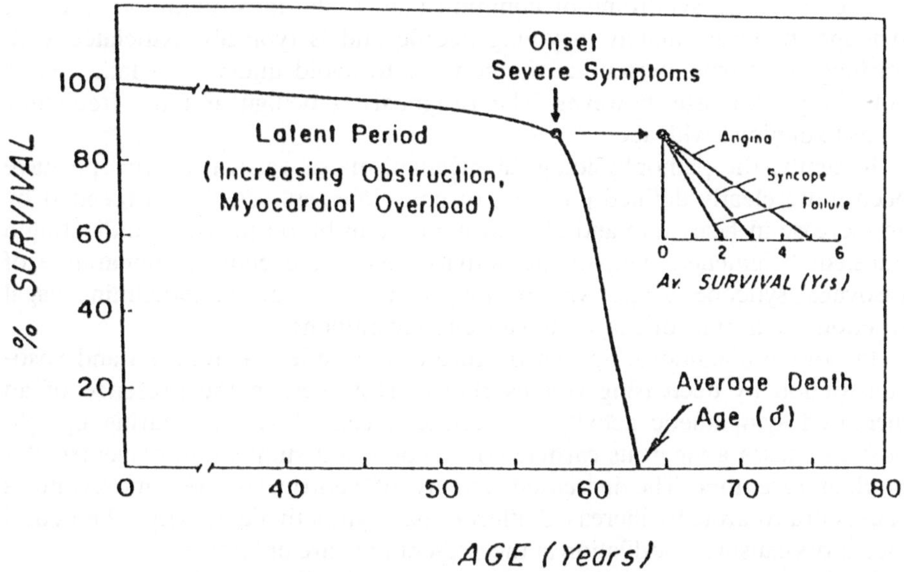

Figure 1. The natural history of aortic stenosis. In the adult patient with acquired aortic valvular stenosis there is a long latent period during which no symptoms occur. During that period survivorship is nearly normal. Once symptoms occur, the survival curve abruptly changes, and mortality is markedly increased. From Ross & Braunwald [3]. Reproduced by permission of the American Heart Association.

congenital or acquired, might account for postural hypotension. Excessive use of diuretics or antihypertensive drugs has always to be considered as a possible cause of syncope in patients receiving these drugs. In geriatric patients, polypharmacy, and in patients with ischemic coronary artery disease, the inappropriate use of nitrates should not be disregarded.

3. *Vagally or neurally mediated syncope*
Neurally mediated syncope is due to activation of mechanoreceptors located in the carotid sinus, the aortic arch, or the ventricles, resulting most commonly in bradycardia (cardioinhibitory type), vasodilation and hypotension (vasodepressor type) or in the combination of both. The pure vasodepressor type of syncope seems to be extremely rare.

For many years, neurally mediated syncope has been recognized in the hypersensitive carotid-sinus syndrome. In this syndrome, slight pressure on the carotid sinus results in episodes of asystole and hypotension. Carotid-sinus massage often reproduces the patient's symptoms. The lesion and its location in this syndrome are unknown at the present time. Some data point to the vasomotor center in the brain stem [5]. However, recent work raised the possibility that minor cardiac pathology at the level of the sinus and atrioventricular node might result in local hypersensitivity to acetylcholine [6].

The vasovagal syndrome or common faint is the most common cause of syncope. It occurs mainly in young people and is typically associated with warning symptoms, permitting the sufferer to avoid injury by a fall. This is not always the case, however. The prognosis is benign and the frequency tends to diminish with age.

Recently, the pathophysiological mechanisms of vasovagal syncope have been more clearly defined [7]. Loss of consciousness usually is preceded by an increase in heart rate and also an increase in blood pressure, indicating a transient heightened sympathetic activity before the clinical appearance of vasovagal syncope. Thus, syncope may result from an overwhelming vagal reaction to a normal or exagerated adrenergic influence.

In normal conditions, upright posture causes reflex tachycardia and vasoconstriction by decreasing venous return. However, in the presence of an increased sympathetic activity, the reduced central volume during upright posture causes a vigorous cardiac contraction and stimulation of ventricular mechanoreceptors. The increased activity of ventricular mechanoreceptors causes bradycardia by increased afferent parasympathetic activity. The neural mediators causing vasodilation at the present time are unknown.

Most young patients with vasovagal syncope don't need any treatment. Explanation of the problem, its mechanism and its predisposing factors to the patient and also the patient's partner will be sufficient. In the older patient, however, vasovagal reactions frequently occur without warning, and not seldom are associated with injury (the so-called malignant vasovagal syndrome). Therapy in those patients is warranted. Atropine usually is ineffective since it doesn't affect vasodilation. Beta 1 receptor blocking drugs [8] and disopyramide [9] have proven successful in a limited number of patients during a limited follow-up period. Beta 1 blocking drugs act by inhibiting the forceful and vigorous left ventricular contraction. The beneficial effect of disopyramide is related to its negative inotropic effect, its potent vagolytic action and probably also to peripheral alpha-adrenergic agonistic effects. In patients in whom cardioinhibitory responses predominate, dual chamber pacing is advocated [10].

4. *Arrhythmias*

(a) *Bradyarrhythmias.* Sick sinus syndrome and high grade atrioventricular block always have to be considered as a possible cause of syncope. In sick sinus syndrome loss of consciousness may be caused either by the brady- or the tachy component of the disease. Not seldom, syncope occurs after termination of a supraventricular tachyarrhythmia in the absence of a reliable escape rhythm, while there is marked suppression of sinus node automaticity because of the arrhythmia. The sick sinus syndrome is associated with a high risk of systemic — mainly cerebral — embolization when atrial tachyarrhythmias are present. Loss of consciousness in these patients might be associated with focal neurologic signs. In a report of 128 patients with sinus node dysfunction, the incidence of syncope varied from 59% in patients presenting with brady-tachy syndrome, to 45% in patients presenting with sinus arrest and/or sinoatrial block, to 33% in patients presenting with only persistent sinus bradycardia [11].

The earliest description of loss of consciousness associated with a slow heart rate probably corresponds to AV conduction dysfunction and is identified as the Morgagni—Stokes—Adams syndrome. The ECG correlate of this syndrome is third-degree AV block. In order to prevent such a catastrophic event, extensive studies on AV conduction were performed in search of early ECG findings that could predict which patients were at risk of developing a high degree of AV block.

Conduction disturbances are a common finding in symptomatic as well as asymptomatic populations. AV nodal conduction is influenced by the autonomic tone. First- and second-degree Wenckebach block are usually benign, and related to autonomic effects. They are very common in athletes and in well trained individuals, as well as in young populations during the sleeping period.

Mobitz type-II second-degree AV block is almost never observed in healthy individuals. It is frequently associated to wide QRS complexes, indicating that the level of conduction block is located below the AV node in the His—Purkinje system.

Patients with trifascicular block [prolonged AV conduction time + (right bundle branch + left anterior hemiblock) (right bundle branch block + left posterior hemiblock)] and patients with alternating left and right bundle branch block are especially considered to be at high risk of developing complete heart block. In patients with bifascicular block, electrophysiologic investigations might, as will be considered later, identify high risk subgroups.

(b) *Tachyarrhythmias.* Both ventricular and supraventricular tachyarrhythmias can cause syncope. The hemodynamic consequences of tachycardias are mainly related to the presence or absence of AV relation, the pattern of ventricular activation, the ventricular rate, and the severity of underlying heart diseases. Obviously, cerebral perfusion will be more impaired in the presence of preexisting cerebrovascular disease.

Atrial fibrillation and atrial flutter uncommonly cause syncope. Atrial fibrillation is an ominous arrhythmia in the presence of preexcitation with a short antegrade refractory period of the accessory pathway and in patients with extensive anterior wall myocardial ischemia, or severe left ventricular dysfunction.

Ventricular tachyarrhythmias cause syncope and may next present as sudden death.

Ventricular tachycardia (VT) and ventricular fibrillation (VF) can occur in the setting of ischemia, either in the acute phase of myocardial infarction or in the post-infarction period. They can be caused by cardiomyopathy or by right ventricular dysplasia. Antiarrhythmic drugs might provoke ventricular tachyarrhythmias by proarrhythmic effects and finally idiopathic ventricular tachycardia might occur in the absence of any detectable heart disease.

In the setting of long QT syndromes, torsades de pointes can cause syncope. Based on sympathetic activity, the long QT syndromes have been divided into two groups.

In Group I (adrenergic-dependent long QT syndrome) torsades de pointes typically occur during episodes of high sympathetic tone (exertion, fear, pain, etc.). This group consists of the idiopathic long QT syndromes (Jervell and Lange-Nielsen and Romano—Ward syndromes, both of which are inherited disorders) and a sporadic type. Group I occurs rarely and is therefore of limited relevance in clinical practice.

In Group II (pause-dependent long QT syndrome), torsades de pointes are precipitated by long ventricular cycles or pauses. They are not provoked by emotional stress or physical exertion and are frequently suppressed by isoproterenol or pacing. This type of arrhythmia can be induced by drugs, mainly by antiarrhythmics (especially class IA agents), diuretics, phenothiazines, and tricyclic antidepressants.

Drug induced hyperthyroidism should be suspected if the QT interval shortens during amiodarone treatment.

Neurologic causes

Excluding epilepsy as a cause of syncope always is a challenging clinical problem. The electroencephalogram and some pointers in the clinical history can direct to the correct diagnosis.

Postural tone is controlled by the reticular formation in the brain stem. Blood supply occurs via the vertebrobasilar system. In the vertebrobasilar syndrome (the so-called drop attack) there is loss of postural tone without loss of consciousness. Furthermore, other neurologic signs usually are present.

In the presence of a significant cerebral artery stenosis, reduction of cardiac output might cause syncope, which usually is associated with focal neurologic signs.

Metabolic causes

Among metabolic disorders causing syncope, the most common are hypoglycemia and hyperventilation.

Hyperventilation usually occurs in overtly anxious individuals. It results in hypocarbia, causing cerebral arterial vasoconstriction and a fall in peripheral vascular resistance. It is of interest that the systemic blood pressure does not fall profoundly, yet patients can lose consciousness, even when recumbent. Hyperventilation is commonly accompanied by tingling of the extremities and perioral paresthesia. Forced hyperventilation may reproduce the symptoms. Hyperventilation followed by a vasovagal reaction can cause syncope. The 'fainting lark' is a technique whereby children induce syncope by hyperventilation followed by a Valsalva manoeuvre. Rebreathing is a common known therapy.

Since the brain is highly dependent upon glucose as a substrate, loss of consciousness can result from hypoglycemia. Generally, the patients are diabetic and are taking insulin or other hypoglycemic agents. The diagnosis is usually made by determining the bloodglucose level during an attack or by the prompt response to therapy with intravenous glucose.

Syncope due to hypoglycemia usually is preceded by premonitory symptoms of adrenergic discharge such as tachycardia, diaphoresis, and pallor. Syncope can last for several minutes. The symptoms that generally accompany hypoglycemic syncope can be muted in patients who are medicated with beta-adrenergic blocking drugs.

AVAILABLE METHODS TO EVALUATE SYNCOPE

1. *Medical history*

The clinical history remains the most important clue to the diagnosis of the cause of syncope. Whenever possible, both the patient and an observer of the episode should be interrogated.

Cardiovascular obstructive syncope tends to occur during exercise. In myxomata, however, syoncope is related to body posture.

The vasovagal syndrome has a typical pattern. Predisposing factors are usually present. They consist of being out of condition, fatigue, fasting, stuffy rooms, emotion, sight of blood and other medical acts, or pain. Warning signs are common and consist of dizziness, nausea, blurring of vision, yawning, pallor, sweating, and a slow deep respiration. Vasovagal syncope very rarely occurs whilst lying down. On recovery, the patient may remain unwell for hours and prolonged unconsciousness, bradycardia, and hypotension may occur.

In *arrhythmic syncope* warning signs usually are absent, occasionally a brief episode of palpitations is mentioned. The patient will be seen to be pallot and if the attack continues for about 16 sec cyanosed. On recovery, the patient has a bright red flush. Flushing immediately after recovery is typically absent in the vasovagal syndrome. Reorientation of the patient rapidly occurs after the syncopal episode, when sinus rhythm is restored. If the attacks are long-lasting, incontinence and epileptiform seizures are likely to occur.

Carotid sinus syndrome has almost no warning. It is commonly assumed that tight collars and head rotation are the triggers of the syncopal episode. This, however, occurs rarely.

As mentioned before, exclusion of *epileptic seizures* remains a challenging clinical problem. As compared to cardiac syncope, pallor is usually absent in epilepsy. There is no episode of pre-syncope and quick orientation is absent. Tonguebiting in cardiac syncope occurs almost never. However, preictal amnesia, convulsions, and urinary or faecal incontinence are able to occur both in epileptic and cardiac syncope, depending on the duration of the attack.

All too often, *cardiogenic orthopedics* are not considered. One should always ask the patient with a bone fracture about the reason why he fell.

Finally, when a patient's first episode of syncope occurs after a *myocardial infarction*, ventricular tachycardia is the most likely mechanism.

Relevant clinical findings, obtainable from the clinical history, are summarized in Table 2.

2. *Physical examination*

Obviously, blood pressure measurement in the supine and standing position is an easy method to diagnose postural hypotension, often secondary to excessive use or intoxication with drugs.

Cardiac obstructive lesions usually produce characteristic cardiac murmurs. In the elderly patient with cardiac failure the murmur of aortic stenosis may be difficult to recognize. Provocative manoeuvres have to be performed when hypertrophic cardiomyopathy is suspected.

Carotid bruits and focal neurological signs should suggest a neurologic cause.

Table 2. Relevant clinical findings in syncope.

1. *Circumstances*: exercise, posture, coughing, micturition, fatigue, fasting, stuffy rooms
2. *Warning signs*: dizziness, nausea, blurring of vision
3. Color of the skin during the attack (pallid or not, immediate flushing after recovery)
4. Duration of unconsciousness
5. Associated findings (tonguebiting, seizures, incontinence)
6. Postsyncopal behaviour (quick orientation versus prolonged symptoms)
7. Previous drug intake (nitrates, antihypertensive agents, hypoglycemic agents, antiarrhythmic drugs)
8. Known underlying cardiac or cerebrovascular disease

Carotid-sinus massage may be very informative. The procedure has to be correctly performed, while precautions are considered (Table 3) [12]. Borderline findings (e.g., 2—3 second pause) during carotid sinus pressure testing should not be overinterpreted. A period of asystole of more than 3 sec is generally regarded as abnormal. However, pauses of 3 to 5 sec are not unfrequently observed in *a*symptomatic elderly patients. In patients with syncope and asymptomatic 3 to 4 sec pauses in response to carotid sinus pressure, carotid sinus hypersensitivity remains a presumptive diagnosis and the indication for pacemaker implantation not always obvious. On the other hand, if carotid sinus massage produces a long pause associated with symptoms, the diagnosis is more convincing. Some asymptomatic patients may become symptomatic only after testing is performed in the upright position.

A subclavian artery lesion has to be suspected if the radial pulse is delayed and attenuated, if the blood pressure difference between left and right arm consistently exceeds 20 mmHg, or when a murmur is present in the supraclavicular fossa. The latter, however, disappears if complete occlusion occurs. When syncope is provoked by exercising the arm ipsilateral to a stenosed subclavian artery, the subclavian steal syndrome also has to be considered.

3. *The resting 12-lead electrocardiogram*

The 12-lead electrocardiogram is definitely helpful in defining the cause of an arrhythmic origin of syncope, when recorded during an acute attack. However, this seldom occurs. Usually, ECG abnormalities in the resting ECG point to underlying cardiac disease (e.g. ischemic heart disease) but seldom lead to an accurate diagnosis of syncope. Preexcitation might suggest a tachyarrhythmic cause. A prolonged QT interval may indicate a congenital or drug induced long QT syndrome and the possibility of 'torsades de pointes'. Chronic bifascicular block can progress to advanced heart block, the incidence, however, is low (about 1% per year). In patients with ischemic heart disease, the incidence seems to be higher. If in these patients syncope occurs, the mechanism may be complete heart block, but also ventricular

Table 3. Rules for Carotid-Sinus Massage (CSM).

Precautions:
1. Try to exclude stenosis of one of the carotid arteries by palpation and auscultation
2. Register the effect of CSM with the help of an electrocardiogram; if not available, listen to the heart with the stethoscope during CSM
3. Realize that the older patient might respond with a long period of cardiac arrest

Proper positioning:
1. Patient flat on his back with his head falling backward (either by placing the physician's left arm or a small pillow under the patient's shoulders)
2. Patient's head turned to the left during massage of the right carotid sinus, and vice versa

Correct massaging:
1. On the bifurcation of the carotid artery, just below the angle of the jaw
2. Massage one side at a time
3. Start by slight pressure (some, especially elderly patients are very sensitive). Thereafter, firm pressure should be applied with a massaging action
4. Do not exceed 5 sec at a time

tachycardia [13]. Sinus bradycardia, first degree or Wenckebach AV block are not uncommon findings in the normal population, especially in athletes. Serious abnormalities (e.g. complete heart block, ventricular tachycardia), may help define a cause of the syncopal episode when recorded in between attacks.

4. *Ambulatory electrocardiographic monitoring*

Prolonged (24 to 48 hours) electrocardiographic monitoring may be useful to document intermittent transient arrhythmias and to correlate symptoms with electrocardiographic findings. If the patient has symptoms of dizziness, presyncope or syncope at the time of an electrocardiographic abnormality (e.g. long sinus pauses, complete heart block, ventricular tachycardia) the diagnosis is readily established. If abnormalities are absent during a syncopal attack, an arrhythmic origin can be refuted. The recording of abnormal rhythms in the absence of symptoms does not necessarily relate to etiology. Many arrhythmias including frequent ventricular ectopy or Wenckebach block are commonly detected in the general population. In young patients, sinus pauses exceeding 1¾ to 2 sec are common during sleep. The incidence and probably also the diagnostic significance of abnormalities is higher in the elderly age group and in patients with structural heart disease [14, 15].

Ector et al. concluded that asymptomatic patients with ventricular pauses of 3 sec or more should be paced [16]. These findings however have been disputed. Mazuz et al. considered such pauses benign and no correlation with symptoms or subsequent sudden death could be identified [17].

Fifty-two patients with pauses greater than 3 sec were retrospectively studied during a relatively short followup period of 1 year. The patients were divided into two groups with a similar incidence of underlying structural heart disease. The duration of the pauses also was similar in both groups. One group was permanently paced. No difference was observed either in mortality or in recurrence of syncope [18].

In the absence of symptoms the significance of long pauses remains questionable. If, on the other hand, sustained ventricular tachycardia, torsades de pointes or high degree AV block are recorded, further investigation and therapy is indicated.

In patients with nonsustained ventricular tachycardia treatment is frequently given unnecessarily. Symptomatic treatment is not allowed, when it may increase mortality instead of reducing it. Nonsustained ventricular tachycardia can be symptomatic, sometimes merely to the patient, but not seldom only to the physican. If only the physician is symptomatic, common sense suggests physician's (re-)education rather than patient's treatment.

5. *Signal-averaged electrocardiography*

The signal-averaged electrocardiogram is a noninvasive method able to analyze the high-frequency, low amplitude components of the terminal portion of the QRS complex. The technique has proven useful to identify patients at risk for ventricular tachycardia, particularly in the setting of ischemic heart disease. Promising results have been reported in patients with syncope in whom ventricular tachycardia may have been the cause of syncope [19, 20]. Results of recent studies can be summarized as follows:

- patients with an abnormal signal averaged electrocardiogram appear to represent a subgroup with a high incidence of inducible ventricular arrhythmias at electrophysiological testing;
- in 89 to 98% of patients with a normal signal averaged electrocardiogram ventricular tachycardia does not occur spontaneously and is neither inducible;
- the test is not reliable in patients with bundle branch block and those dependent on ventricular pacing at the time of the recording.

6. *Exercise stress testing*

Exercise stress testing may contribute important diagnostic information in selected patients with syncope, particularly in patients with symptoms related to exertion. During and after stress testing different abnormalities can be observed:

- exercise-induced hypotension,
- abnormal chronotropic response suggesting sinus node dysfunction,
- exercise-induced AV block at low rates indicating subnodal disease,
- supraventricular or ventricular tachyarrhythmias, whether or not preceded by ischemic ST-T abnormalities or hypotension,
- occurrence of post-exercise asystole,
- vasomotor instability with hypotension at the end of exercise.

Exercise testing enables identification of a subgroup of patients being at extremely high risk for malignant syncope and subsequent sudden death [21].

When compared to Holter monitoring, the diagnostic power of exercise testing is considered to be inferior in patients with unexplained syncope. However, it has to be stressed that both tests, while addressing a different arrhythmia substrate, rather than being considered as competitive, are in fact complementary.

7. *Upright tilt testing*

As described earlier, vasovagal episodes have a rather typical anamnesis consisting of predisposing factors, warning signs and prolonged symptoms on recovery. However, especially in older patients, an atypical clinical pattern with absent warning signs, might result in injury by fall (malignant vasovagal syndrome, being one of the causes of 'orthopedic' syncope).

Until recently, specific provocative tests to confirm the diagnosis of vasovagal syncope were not available. In the vasovagal syndrome, loss of consciousness is preceded by an increase in sympathetic tone. Normal vascular responses (baroreceptor stimulation) to maintain blood pressure in the standing position are altered. In fact, counterproductive reflexes, due to stimulation of ventricular mechanoreceptors increase afferent parasympathetic activity and cause bradycardia. In the vasovagal syndrome, syncope can be sometimes provoked using an upright tilt test. Patients are tilted up at an angle of 60° for 10 min. The electrocardiogram and the blood pressure are monitored. In order to increase sympathetic activity, isoproterenol is infused at a rate of 1 μg/min for 10 min. The test is repeated with graded infusions up to 5 μg isoproterenol per minute. Finally, the test is terminated after completion of 5 μg/min of isoproterenol, when intolerance to isoproterenol develops, when the heart rate increases to 150 beats/min or when syncope occurs. Results of the test appear to be highly reproducible.

Very recently, tilt testing has been used in conjunction with electrophysiological testing to assess effects of body posture on electrical properties of the heart, both during or prior to the development of tachycardias [22].

Finally, the test can be used to assess the efficacy of pharmacological or non pharmacological therapy in patients with neurally mediated syncope. The sensitivity and specificity of upright tilting, however, should be further defined by more extensive investigation.

8. *Electrophysiological testing*

In patients with syncope and suspected cardiac arrhythmias, undetected by noninvasive evaluation, electrophysiological studies can be performed with a very low incidence of complications. Electrophysiological testing involves:

(1) assessment of sinus node function, and
(2) AV nodal and His–Purkinje conduction,
(3) programmed extrastimulus testing to induce supraventricular and ven-

tricular tachyarrhythmias and to assess anomalous conduction over an accessory pathway.

If syncope occurred during antiarrhythmic drug treatment, EP studies should be performed both on and off medication. The diagnostic yield of EP testing in patients with unexplained syncope has been differently appreciated. In different studies the incidence of electrophysiological abnormalities ranged from 12 to 70% [23, 24]. Disparity in these results can be explained in several ways:

(1) the diagnostic yield will become higher, the more patients with underlying heart disease are studied;
(2) the diagnostic end-point of EP testing has been differently appreciated. When ventricular tachyarrhythmias are induced by programmed stimulation, sustained monomorphic ventricular tachycardia has a much more powerful diagnostic value as compared to non sustained or polymorphic ventricular tachycardias;
(3) the arrhythmia induction protocol. Technology 'advanced' that much that in any human being ventricular fibrillation can be induced.

In clinical practice it is desirable that patients with unexplained syncope can be stratified into subgroups with a high or low probability of having an electrophysiological abnormality that is likely to be causally related to syncope. As a general rule in medicine, risk stratification has to be performed on clinical variables, ideally readily obtained by every physician and not requiring sophisticated and complicated investigations (Table 4) [25].

(a) *Sinus node dysfunction.* Sinus node dysfunction is a common cause of arrhythmogenic syncope causing pacemaker implantation. Its prevalence increases in the older age group. Electrophysiologic testing of sinus node dysfunction does not make an important contribution to the care of patients with sinus node disorders [26] in terms of the bradyarrhythmia. However, it

Table 4. Findings associated with abnormal and normal electrophysiological outcomes in patients undergoing testing for unexplained syncope.

Abnormal findings	Normal findings
– Presence of underlying structural heart disease	– Lack of underlying heart disease
– Prior myocardial infarction	
– Poor ventricular function	– Normal ventricular function
– Advanced congestive failure	
– Atrial fibrillation	
– Older age	– Younger age
– Q waves on ECG	– Normal ECG
– Bifascicular block	
– Late potentials identified by signal averaged ECG	– Late potentials lacking

may be very important to prevent the atrial tachyarrhythmias. While an electrophysiological study can help document the presence of sinus node dysfunction, it generally cannot help to assess whether such dysfunction is causing symptoms. The reverse is also true. Management of those patients is more appropriately conducted using clinical and electrocardiographic data derived from the 12 lead ECG and mainly from ambulatory Holter monitoring. Different parameters are commonly used to assess sinus node dysfunction:

(1) the sinus node recovery time (SNRT), considered to be abnormal when exceeding 1500 msec;
(2) the corrected sinus node recovery time (CSNRT) being calculated by substracting the cycle length of the patient's sinus rhythm in msec from the SNRT (upper normal limit: 550 msec);
(3) the sino-atrial conduction time (SACT) reflecting the time it takes an impulse originating in the sinus node to travel through the perinodal tissue into the atria (normal value 50—125 msec);
(4) failure of the heart rate to increase above 90 beats per minute after intravenous injection of Atropine.

Permanent pacing in patients with unexplained syncope who had an abnormal sinus node function test has not been uniformly successful. While there is some variation from laboratory to laboratory, the sensitivity and specificity for both SNRT and SACT tests are generally only approximately 70%, making these tests less than ideal.

Those tests, at the present time, are not indicated in asymptomatic patients with bradyarrhythmias. In symptomatic patients with documented bradyarrhythmias on Holter, cardiac pacing will not be preceded by EP evaluation of sinus node function. The only true indication for performing EP studies to assess sinus node function is in patients who present with symptoms (especially syncope suggestive of bradyarrhythmias on clinical grounds) and in whom no bradycardia is documented during cardiac monitoring.

(b) *AV nodal and His-Purkinje conduction.* Evaluation of AV conduction during electrophysiological testing includes determination of the baseline AV node to His bundel conduction (A-H interval), the His bundle to ventricular conduction time (H-V interval) and the duration of the His bundle deflection. AV nodal and His—Purkinje refractory periods are determined by the extrastimulus technique. Incremental atrial pacing is used to assess AV conduction, by determining the atrial rate at which the onset of Wenckebach and 2 to 1 AV block occurs, and by determining the level of the induced block (supra-, intra- or infra-Hisian). In general, distal rather than proximal conduction disease is of greater importance and more likely to be causally related to a syncopal event.

AV nodal conduction is influenced by autonomic tone and first and Wenckebach degree AV block are usually benign and related to autonomic effects. Pacing is only required if the resultant bradycardia produces symptoms. When possible, drugs that slow AV conduction have to be discontinued.

Intermittent high-grade AV block (Adams—Stokes attacks) with normal conduction between episodes is a common cause of syncope. Usually it is detected by Holter monitoring and only rarely accounts for unexplained syncope.

The significance of a prolonged HV interval during EP testing remains controversial. In patients with preexisting chronic bifascicular block, the risk for developing complete infranodal block seems to be related to the degree of HV prolongation. Scheinman et al. [27] found that progression to complete heart block occurred in 12% of patients with HV $\geqslant$ 70 msec, compared with 2% of normals and 4% of those with HV < 70 mesc. In a 3 year follow-up, 24% of patients with HV $\geqslant$ 100 msec developed complete heart block. It has to be stressed that since AV block might be intermittent, the finding of a normal HV interval doesn't exclude conduction disease.

(c) *Supraventricular tachycardias.* In patients with supraventricular tachycardia, symptoms such as palpitations or mild dizziness commonly occur. Syncope, however, is not a prominent symptom. If syncope occurs it is related to tachycardia-induced hypotension. Hypotension more likely occurs at the onset of the tachycardia, in the standing position and in the patient with marked ventricular dysfunction or critical valvular stenosis.

In the presence of preexcitation with a short antegrade refractory period of the accessory pathway and in the setting of extensive anterior wall myocardial ischemia, atrial fibrillation not seldom causes loss of consciousness due to marked hypotension.

When a supraventricular tachycardia is induced during EP testing a causal relationship for a syncopal episode only can be considered if the tachycardia is very rapid (> 180/min) or when tachycardia related hypotension develops. If the induced tachycardia is slow and not associated with hemodynamic disturbances, other causes of syncope have to be considered.

In the same patient, more than one tachycardia may be induced. The tachycardia that reproduces the clinical findings generally is considered the more specific finding.

(d) *Ventricular tachycardia.* Ventricular tachycardia and ventricular fibrillation are the most common arrhythmias induced by programmed electrical stimulation in patients with unexplained syncope. Induction of those arrhythmias is easier in patients with underlying heart disease. Clinically nonrelevant arrhythmias can be provoked by an aggressive stimulation protocol using more than two extrastimuli. Sustained monomorphic ventricular tachycardia induced by one or two ventricular extrastimuli is a relatively specific finding [28]. Non clinical forms of tachycardia tend to be nonsustained, polymorphic and fast [29, 30].

In patients with bundle branch block and unexplained syncope, the incidence of ventricular tachycardia initiated by programmed stimulation is increased. In those patients, especially when HV prolongation is present, syncope either can be caused by ventricular tachycardia or heart block [31].

Finally, electrophysiological testing has been suggested to guide initial and subsequent antiarrhythmic drug therapy. Serial testing however is cost-ineffective and can have considerable psychological consequences for the patient because of the multiple studies performed and prolonged and repeated hospitalization. An algorithm based on simple parameters, readily obtainable by medical history, recently has been developed in order to facilitate decision taking in patients with ventricular arrhythmias post myocardial infarction [32].

PROGNOSIS OF SYNCOPE

The prognosis of syncope can be defined in terms of total mortality, sudden death and recurrences. As far as total mortality and sudden death is concerned, it was proven to be extremely relevant to stratify patients according to the etiology of syncope. Patients with a cardiac origin of syncope significantly differed in outcome as compared to patients with a noncardiac cause or patients in whom an etiology was undetermined by noninvasive diagnostic tests. At 12 months, mortality was 30% in patients with cardiac syncope, 12% in noncardiovascular syncope and 6.4% in patients with an unknown cause. The one year incidence of sudden death was 24% in cardiac, 4% in noncardiac and 3% in patients with unexplained syncope [1, 33]. In other words, the prognosis of cardiac syncope is poor. Noncardiac syncope and syncope of unknown origin have a similar good prognosis.

Recurrences of syncope occur commonly (35% recurrence rate within 5 years). There is no difference in recurrence rate in cardiac, noncardiac or unexplained syncope. The majority of recurrences occur within the first two years. Recurrences are not predictive of total mortality or sudden death. At the time being, one is unable to device any strategies able to predict recurrences.

The prognosis of ventricular tachycardia is highly dependent on its etiology (Table 5). Idiopathic VT and VT occurring in the setting of right ventricular dysplasia both have a very low mortality. However, the arrhythmias are very likely to recur. VT and VF caused by ischemic coronary artery disease have a high mortality rate, a high incidence of sudden death and a high recurrence rate [34].

In daily clincal practice, life threatening syncope due to ventricular tachyarrhythmias seems merely to be confined to patients with a previous myocardial infarction. Even in this subgroup, risk stratification can identify patients at high risk for sudden death despite appropriate antiarrhythmic drug treatment. By using 5 variables, easily to obtain by clinical history, a subgroup of patients can be identified in whom antiarrhythmic drug treat-

Table 5. Relationship between etiology of VT and VF to mortality and recurrences.

	Previous MI		RVD	Idiopathic	
	VT	VF	VT	VT	VF
Total number of patients	79	37	11	52	6
Mortality	15 (19%)	13 (35%)	1 (19%)	1[a] (2%)	0
Sudden death	5 (6%)	5 (14%)	0	0	0
Recurrences arrhythmia	24 (30%)	6 (19%)	6 (67%)	15 (23%)	2 (30%)
Follow-up (months)	26	22	39	86	39

[a] Mortality: cancer.
RVD, Right ventricular dysplasia; VF, Ventricular fibrillation; VT, Ventricular tachycardia

ment results in an unacceptable sudden death rate of >10% in a 2 year follow-up period [32, 35, 36]. In those patients nonpharmacologic therapy has to be considered (antiarrhythmic surgery, chemical or electrical ablation, antitachycardia pacemakers, automatic implantable cardioverters and even sophisticated devices such as the combination of an antitachycardia pacemaker and an implantable defibrillator in the same unit). The high risk subgroup consists of patients

(1) in whom the initial episode of the arrhythmia was poorly tolerated resulting in syncope or sudden death requiring immediate resuscitation,
(2) who, outside the attack of arrhythmia have symptoms of dyspnea (NYHA class III),
(3) who had exercise induced monomorphic sustained ventricular tachycardia,
(4) who suffered from multiple myocardial infarctions,
(5) in whom the interval between the attack of ventricular tachycardia/fibrillation and the onset of the last episode of acute myocardial infarction was more than 3 days and less than 2 months.

The risk of sudden death in asymptomatic post MI patients with frequent ventricular premature beats or nonsustained ventricular tachycardia has proven to be small. Recent results of the CAST-trial [37] indicate that antiarrhythmic treatment with flecainide and encainide increases mortality in those patients. The results of this trial have almost been predicted by Brugada et al. [38]. In order to induce sustained monomorphic ventricular tachycardia, 160 patients who recovered from myocardial infarction were investigated by programmed electrical stimulation. The patients were divided into 4 groups: Group I consisted of patients with no ventricular arrhythmias, Group II patients had attacks of nonsustained ventricular tachycardia, Group III had previously been resuscitated from ventricular fibrillation, whereas Group IV patients suffered from sustained ventricular tachycardia.

Sustained monomorphic ventricular tachycardia could be induced in nearly all (93%) of Group IV patients and in more than 40% of patients of Groups I and II. In the latter, however, the number of extrastimuli was higher, the tachycardia cycle length shorter and by consequence the tachycardia rate faster. This implies the presence of a reentry circuit in about half of the patients of Groups I and II. However, the electrophysiological properties of this circuit are not prone to induction of malignant arrhythmias. Since both the number of extrastimuli and the rate of induced tachycardia are higher, the refractory period of this circuit has to be shorter and the conduction velocity faster. In those patients antiarrhythmic agents, by increasing the refractory period and by slowing conduction velocity, can induce proarrhythmic activity.

WHAT IS THE DIAGNOSTIC YIELD OF 'DIAGNOSTIC' TESTS?

The diagnostic yield of different tests (Table 6) was studied by Kapoor in 433 patients admitted for syncope. In 179 patients a cause couldn't be identified. Clinical history and physical examination were able to assign the cause in 140 patients, the ECG in 30 and Holter recordings in 54. Cerebral angiography was diagnostic in 2 patients with the subclavian steal syndrome. The electroencephalogram, although performed in all 433 patients, was only helpful in 2. Those patients had other neurologic signs or symptoms, suggesting a neurologic origin of syncope on the basis of clinical history alone.

HOW TO EVALUATE THE PATIENTS WITH SYNCOPE?

Syncope is not only unfair. It attacks frequently and conceals rapidly afterwards. It has a wide variety of causes, each cause having a driving pressure to increase the amount of sophisticated, costly diagnostic tests. In this situation, only the most uncommon of senses makes sense: common sense. Physical examination and particularly the clinical history are diagnostic in one third of

Table 6. Diagnostic yield of tests commonly used in syncope.

Total number of patients	433	Stress testing	2
No diagnosis	179	CSM	5
History and physical examination	140	Catheterization	11
ECG	30	Cerebral angiography	2
Holter	54	EEG	2
EPS	7	Autopsy (dissecting aortic aneurysm)	1

Source: Kapoor et al. [1]

patients with syncope. When not diagnostic, they commonly enable goal directed specific testing. The 12-lead ECG, carotid sinus massage and Holter monitoring are inexpensive and suggest or confirm a diagnosis in one fourth of the patients. Laboratory testing should be kept as simple as possible (complete blood count, glucose, blood urea nitrogen, creatinine and electrolytes). If cardiac obstructive lesions are suspected during physical examination, echo-Doppler studies and/or invasive hemodynamic and angiographic evaluation are indicated. Electrophysiological testing usually is not indicated when symptomatic episodes have been recorded otherwise. However, in the absence of such recordings EP testing may be indicated. Reasons to perform such a study at our institution are:

(1) when a cardiac origin of syncope is suspected by clinical history,
(2) when a patient's first episode of syncope occurs after myocardial infarction,
(3) in elderly patients,
(4) when recurrences regularly occur,
(5) when ECG abnormalities, although not diagnostic, suggest an arrhythmic cause,
(6) in patients with serious injury related to falls.

Although a normal EP study can exclude an arrhythmic cause of syncope (negative predictive accuracy > 95%), and reassures both the patient and the physician, it always has to be considered some diagnoses made in the EP laboratory are only presumptive. Such diagnoses may form the basis for therapeutic trials that may need to be reevaluated when symptoms recur. The diagnostic yield of the electroencephalogram (EEG) is very limited. In the absence of symptoms or signs suggesting a neurologic cause of syncope, the EEG is seldom helpful. The use of upward tilting test should be expanded to increase diagnostic information. Such information doesn't offer equal therapeutic benefits in all subgroups of patients with neurally mediated syncope.

Finally, for the trully recurrent syncope of unknown cause after extensive non-invase and invasive investigation, a last remedy is always available. In the words of Dr. Furman: 'Give him 1000 dollars and ask him to look for another physician'. Certainly, syncope of unknown cause may represent a problem to many physicians. Not to Dr. Furman who, by the way, never followed his own advise, but neither to all those physicians who 'experience' problems not as problems, but only as challenges.

Salient features

The whole is simpler than the sum of its parts (W. Gibbs).

The cost of living is going up and the chance of living is going down (F. Wilson).

Before ordering a test decide what you will do, if:

(1) the test is positive,

(2) the test is negative.

If both answers are the same, don't do the test.

It is not easy to explain why common sense is the most uncommon of all senses.

REFERENCES

1. Kapoor WN, Karpe M, Wieand S, Peterson JR, Levey GS (1983) A prospective evaluation and follow-up of patients with syncope. *N Engl J Med* 309: 197
2. Medina RP, Dreifus LS (1983) Syncope. *Curr Probl Cardiol* 8(8)
3. Ross J Jr, Braunwald E (1986) Aortic stenosis. *Circulation* 38 (suppl V): 61
4. Paulus WJ, Nellens P, Heyndrickx G, Andries E (1985) The influence of chronic amiodarone treatment on left ventricular relaxation, rest and exercise hemodynamics in patients with hypertrophic cardiomyopathy. *Circulation* 72(3): 446
5. Morley CA, Sutton P (1984) Carotid sinus syndrome (editorial review). *Int J Cardiol* 6: 287
6. Kaseda S, Zipes DP (1988) Vagal denervation of canine sinus and atrioventricular nodes creates supersensitivity to Acetylcholine. *J Am Coll Cardiol* 11: 39A
7. Abboud FM (1989) Ventricular syncope: Is the heart a sensory organ? *N Engl J Med* 320: 390
8. Goldenberg IF, Almquist A, Dunbar D (1987) Prevention of neurally-mediated syncope by selective beta-1 adrenoceptor blockade (abstract). *Circulation* 76: 133
9. Milstein S, Buetikofer J, Lesser J (1987) Disopyramide reversal of induced hypotension bradycardia in neurally mediated syncope (abstract). *Circulation* 76: 175
10. Morley CA, Sutton R (1984) Carotid sinus syndrome (editorial review). *Int J Cardiol* 6: 287
11. Simonsen E, Straede NJ, Lyager NB (1980) Sinus node dysfunction in 128 patients: A retrospective study with follow-up. *Acta Med Scand* 208: 343
12. Wellens HJJ, Brugada P, Bär F (1986) p 128 ff in: Kulbertus HE (ed), *Diagnosis and Treatment of the Regular Tachycardia with a Narrow QRS Complex*. Churchill-Livingstone
13. Morady F, Higgins J, Peters RW, Schwartz AB, Shen EN, Bhandari A, Scheinman MM, Sauvé MJ (1984) Electrophysiologic testing in bundle branch block and unexplained syncope. *Am J Cardiol* 54: 587
14. Johansson BW (1977) Long-term ECG in ambulatory clinical practice. *Eur J Cardiol* 5: 39
15. Winkle RA (980) Ambulatory electrocardiography. *Mod Conc Cardiovasc Dis* 49: 7
16. Ector H, Rolies L, De Geest H (1983) Dynamic electrocardiography and ventricular pauses of 3 seconds and more: Etiology and therapeutic implications. *PACE* 6: 548
17. Mazuz M, Friedman HS (1983) Significance of prolonged electrocardiographic pauses in sino-atrial disease: Sick Sinus Syndrome. *Am J Cardiol* 52: 485
18. Hilgard J, Ezri MD, Denes P (1985) Significance of ventricular pauses of three seconds or more detected on twenty-four-hour Holter recordings. *Am J Cardiol* 55: 1005
19. Winters SL, Steward D, Gomes JA (1987) Signal averaging of the surface QRS complex predicts inducibility of ventricular tachycardia in patients with syncope of unknown origin: A prospective study. *J Am Coll Cardiol* 10: 775
20. Kuchar DL, Thorburn CW, Sammel NL (1986) Signal averaged electrocardiogram for evaluation of recurrent syncope. *Am J Cardiol* 58: 1014

21. Rodriguez LM, Waleffe A, Brugada P, Dehareng A, Lezaun R, Sternick E, Kulbertus HE (1990) Exercise-induced sustained sympathetic ventricular tachycardia: Incidence, clinical, angiographic and electrophysiologic characteristics. *Eur Heart J* 11: 225
22. Hammill SC, Homes DR, Wood DL (1984) Electrophysiological testing in the upright position: Improved evaluation of patients with rhythm disturbances using a tilt table. *J Am Coll Cardiol* 4: 65
23. Hess DS, Morady F, Scheinmann MM (1982) Electrophysiologic testing in the evaluation of patients with syncope of undermined origin. *Am J Cardiol* 50: 1309
24. DiMarco JP, Garan H, Ruskin JN (1983) Approach to the patient with syncope of unknown cause. *Mod Conc Cardiovasc Dis* 52: 11
25. Brooks R, Garan H, Ruskin JN (1990) Evaluation of the patient with unexplained syncope, p 647 ff in: Zipes and Jalife (eds), *Cardiac Electrophysiology from Cell to Bedside*. WB Saunders
26. Breithardt G, Seipel L, Loogen F (1977) Sinus node recovery time and calculated sinoatrial conduction times in normal subjects and patients with sinus node dysfunction. *Circulation* 56: 43
27. Scheinman MM, Peters RW, Sauve MJ, Desai J, Abbott JA, Cogan J, Wohl B, Williams K (1982) Value of the HQ-interval in patients with bundle branch block and the role of prophylactic permanent pacing. *Am J Cardiol* 50: 1316.
28. Livelli FD, Bigger J Th, Reiffel JA, Gang ES, Patton JN, Noethling PM, Rolnitzky LM, Glikich JI (1982) Response to programmed ventricular stimulation: Sensitivity, specificity and relation to heart disease. *Am J Cardiol* 50: 452
29. Morady F, Shapiro W, Shen E, Sung RJ, Scheinman MM (1984) Programmed ventricular stimulation in patients without spontaneous ventricular tachycardia. *Am Heart J* 107: 875
30. Brugada P, Green M, Abdolah H, Wellens HJJ (1984) Significance of ventricular arrhythmias initiated by programmed ventricular stimulation: The importance of the type of ventricular arrhythmia induced and the number of premature stimuli required. *Circulation* 69: 87
31. Morady F, Higgins J, Peters RW, Schwartz AB, Shen EN, Bhandari A, Scheinman MM, Sauvé J (1984) Results of electrophysiologic testing in patients with bundle branch block and unexplained syncope. *Am J Cardiol* 54: 587
32. Brugada P, Talajic M, Smeets J, Mulleneers R, Wellens HJJ (1989) The value of the clinical history to assess prognosis of patients with ventricular tachycardia or ventricular fibrillation after myocardial infarction. *Eur Heart J* 10: 747
33. Kapoor WN, Peterson J, Wieand HS (1987) Diagnostic and prognostic implications of recurrences in patients with syncope. *Am J Med* 83: 700
34. Trappe HJ, Brugada P, Talajic M, Della Bella P, Lezaun R, Mulleneers R, Wellens HJJ (1988) Prognosis of patients with ventricular tachycardia or fibrillation: Role of the underlying etiology. *J Am Coll Cardiol* 12: 166
35. Brugada P, Lemery R, Talajic M, Della Bella P, Wellens HJJ (1987) Treatment of patients with ventricular tachycardia or fibrillation: First lessons from the 'Parallel study', p 457 ff in: Brugada P, Wellens HJJ (eds), *Cardiac Arrhythmias: Where to go from here?* Mount Kisco NY: Futura Publishing Company
36. Brugada P, Wellens HJJ (1986) Need and design of a prospective study to assess the value of different strategic approaches for management of ventricular tachycardia or ventricular fibrillation. *Am J Cardiol* 57: 1180
37. The Cardiac Arrhythmia Suppression Trial (CAST) Investigators (1989) Preliminary report: Effect of encainide and flecainide on mortality in a randomized trial of arrhythmia suppression after myocardial infarction. *N Engl J Med* 321: 406
38. Brugada P, Waldecker B, Kersschot I, Zehender M, Wellens HJJ (1986) Ventricular arrhythmias initiated in four groups of patients with healed myocardial infarction. *J Am Coll Cardiol* 8: 1035

2. Pseudobradyarrhythmias

PEDRO BRUGADA, SINAN GÜRSOY, JACOB ATIÉ,
MARC GOETHALS, GÜNTER STEURER, HILDE WILLEMS,
JOSEP BRUGADA & ERIK ANDRIES

INTRODUCTION

An inappropriately slow heart rate is known as a bradyarrhythmia. Bradyarrhythmias may occur because of disorders of impulse formation, disorders of impulse conduction, or a combination of both. Common causes of bradyarrhythmias are disorders in sinus node function and retarded or blocked conduction in the atrio-ventricular node or subnodal conduction system.

These disorders are relatively easy treated by means of implantable pacemakers. The easiness of pacemaker implantation may have decreased a bit the interest in the diagnosis of the underlying mechanisms of bradyarrhythmias. Some patients, however, may unnecessarily receive a pacemaker because of a wrong diagnosis of bradyarrhythmia. Actually, there are a series of disorders of impulse formation and conduction which are not true bradyarrhythmias. They simulate a bradyarrhythmia and can be called 'pseudobradyarrhythmias'.

PSEUDOBRADYARRHYTHMIAS SIMULATING DISORDERS IN IMPULSE FORMATION

One of the most common causes of a wrong diagnosis of sick sinus syndrome is atrial premature beats not conducted to the ventricles. The diagnosis of these atrial premature beats is easily made if they are conducted anterogradedly. However, when they are not conducted, and particularly when only single channel electrocardiograms are available, the diagnosis may be extremely difficult.

Atrial premature beats may invade the sinus node and reset the sinus cycle. That results in a pause. When these extrasystoles occur in a bigeminal form, sinus bradycardia is simulated. The only way to be sure of the correct diagnosis is to record multiple electrocardiographic leads. It has to be

E. Andries, P. Brugada & R. Stroobandt (eds.), How to face 'the faces' of cardiac pacing, 23–28.

realized, however, that a correct diagnosis may require endocavitary recordings in exceptional cases.

A similar phenomenon occurs after ventricular prematue beats retrogradedly conducted to the atria. The sinus node is reset because of retrograde invasion and a pause is produced. This should rarely, if ever, result in a wrong diagnosis of sinus bradycardia and sinus node disease.

Extrasystoles can originate anywhere in the heart. The sinus node is not an exception. When sinus node extrasystoles are conducted to the atria through the sino-atrial junction, the atria are activated and diagnosis should not be more difficult than for regular atrial extrasystoles. Typical for sinus node extrasystoles are a P wave morphology practically identical to the sinus P wave, and a return cycle which is practically identical to the sinus cycle length (because of the absence of any retrograde and anterograde conduction delay in the sino-atrial junction). It is theoretically and practically possible, however, to have sinus node extrasystoles which are blocked in the sino-atrial junction and do not excite the atria. In this situation, the sinus node extrasystole is not visible on the surface electrocardiogram because it does not result in a P wave. The only observation is a doubling of the sinus node cycle which can be mistakenly diagnosed as 2:1 sino-atrial block. There is no reliable means to make the diagnosis of sinus node extrasystoles blocked in the sino-atrial junction. They can only be suspected if the same patient shows conducted sinus node extrasystoles with a variable coupling. It has to be stated, however, that we do not have any effective treatment for this condition. If the extrasystoles give raise to frequent episodes of sinus node pauses, the only option to control the heart rate will still be a pacemaker.

PSEUDOBRADYARRHYTHMIAS SIMULATING DISORDERS IN IMPULSE CONDUCTION

False disorders in impulse conduction are simulated by three conditions: (1) the presence of dual atrio-ventricular nodal pathways, (2) His bundle extrasystoles, and (3) ventricular premature beats.

1. *Dual atrio-ventricular nodal pathways.* The atrio-ventricular node is a very complex structure. Longitudinal dissociation into multiple functional pathways seems to be the rule, rather than the exception. In the electrophysiology laboratory, longitudinal dissociation seems to occur over two functionally different pathways. In this situation, the atrio-ventricular node consists of two separate pathways, one with normal, relatively fast conduction, and a long refractory period, and a second conducting pathway with slow conduction and a short refractory period. During sinus rhythm, conduction occurs over the fast pathway and the PR interval is normal. However, when the heart rate increases, conduction may block in the fast pathway and shift to conduction over the slow pathway. The PR interval prolongs abruptly

and remains long (Figure 1). Depending upon the duration of the refractory period of the fast pathway, condition may shift to the slow pathway at relatively slow atrial rates. The electrocardiogram shows 1:1 atrio-ventricular conduction with a long PR interval and simulates a extreme first-degree atrio-ventricular conduction delay. This phenomenon may be enhanced by drugs affecting atrio-ventricular nodal conduction such as digitalis, beta-blockers, or calcium channel blockers. In reality, what we observe is an example of pseudo-first-degree atrio-ventricular block. A shift to slow pathway conduction can also result in a Wenchebach sequence of conduction in the slow pathway because of retrograde invasion of this pathway. A pseudo-second-degree atrio-ventricular block is simulated in that case. A pseudo-third-degree atrio-ventricular block can occur when more than two pathways are present in the atrio-ventricular node. Concealed invasion of the pathways by the others may result in very complex conduction patterns and blocked P waves.

A shift from fast to slow atrio-ventricular nodal conduction can also occur after extrasystoles originating in the bundle of His or in the ventricle and invading retrogradely the atrio-ventricular nodal pathways (Figure 2). Dual atrio-ventricular nodal pathways present clinically most commonly as intra-nodal reentrant tachycardia. It has to be realized that sometimes conduction to the ventricles may occur in a 2:1 fashion and simulate a bradyarrhythmia or a combined normal rhythm with conduction disturbances (Figure 3).

2. *His bundle extrasystoles and ventricular premature beats.* Extrasystoles originating in the conduction system of the heart and ventricular extrasystoles can be conducted retrogradedly in the conduction system but fail, to be conducted retrogradedly to the atria. In this situation, a prolongation of the

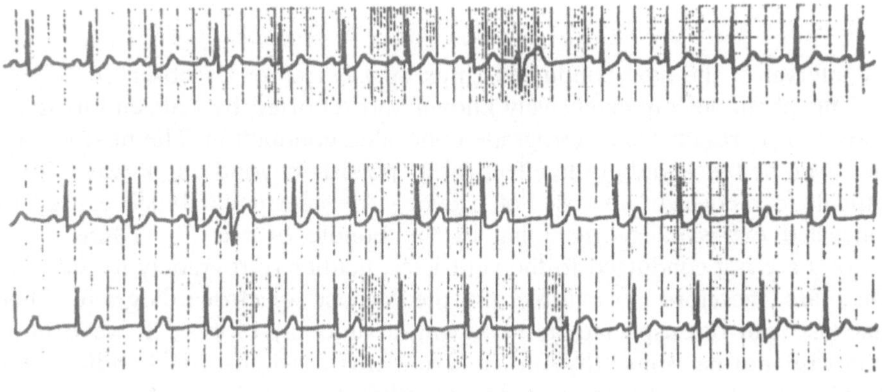

Figure 1. Shift from fast to slow AV nodal pathway conduction after a ventricular premature beat.

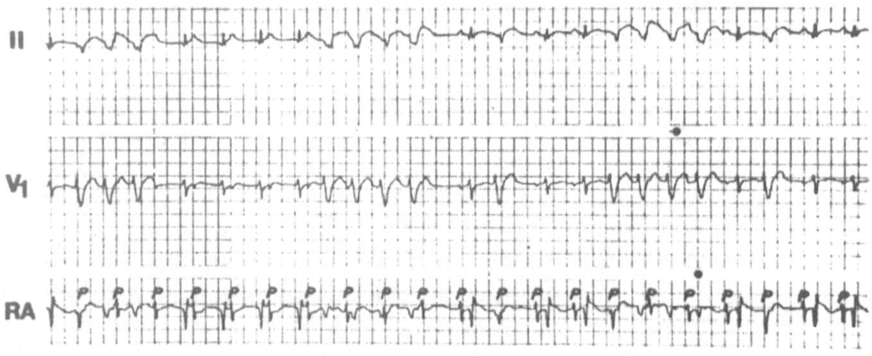

Figure 2. Shift from fast to slow AV nodal pathway conduction during ventricular tachycardia. RA = right atrial electrogram.

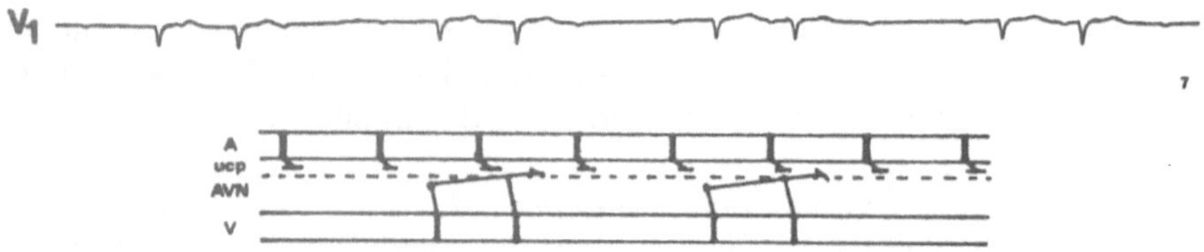

Figure 3. Intranodal reentry producing pseudo-AV block because of retrograde invasion of the upper common AV nodal pathway (UCP).

PR interval of the next conducted sinus beat is frequently observed (Figure 4). This phenomenon, extensively known and reported by Langendorf many years ago [1] results from retrograde concealed conduction. The next P wave may even be blocked, but the effect rarely persists beyond one P wave. More difficult to recognize are cases of pseudo-atrio-ventricular block caused by concealed extrasytoles originating in the bundle of His [2]. These extrasystoles are not conducted to the ventricles, neither retrogradely to the atria. They are therefore not visible on the surface electrocardiogram. Their diagnosis can be suspected by observing a suddenly blocked P wave in an otherwise normal sinus rhythm, in a patient in whom His bundle extrasystoles are observed at other times. Confirmation of the diagnosis requires an intracavitary electrogram of the bundle of His, showing His bundle extrasytoles not conducted to the ventricles neither to the atria.

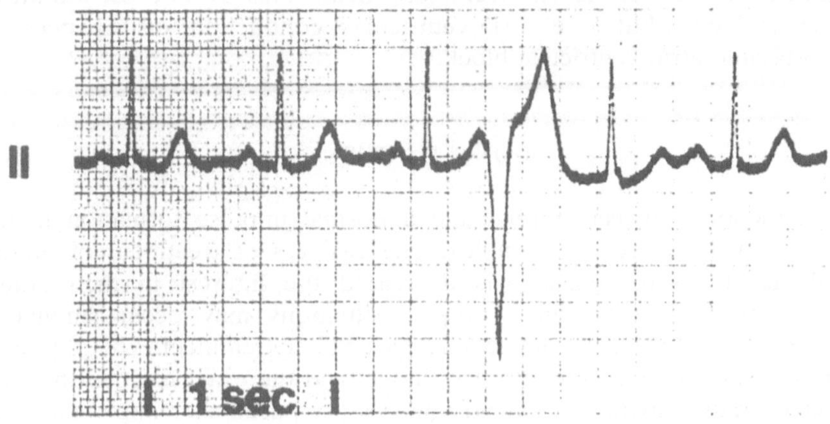

Figure 4. A ventricular premature beat not retrogradedly conducted to the atrium produces prolongation of the next PR interval because of retrograde concealed conduction in the atrio-ventricular conduction system. The P wave is at the T wave of the premature beat.

PSEUDO-PSEUDOBRADYARRHYTHMIAS

Recognition of the exact cause of an arrhythmia requires an open mind. The previous discussion should not lead to a wrong diagnosis of pseudo-brady-arrhythmia when a true bradyarrhythmia is present. There are, indeed, at least two situations in which a true bradyarrhythmia can simulate a pseudo-bradyarrhythmia. We will call these situations 'pseudo-pseudobradyarrhyth-mias'.

1. *Pseudo-pseudo sinoatrial disease.* Atrial premature beats reset the sinus node. Because of the anterograde and retrograde conduction time from the site of origin of the extrasystole to the sinus node and back to the atrium, the pause is longer than the spontaneous sinus node cycle length. There are, however, limits to this situation. A progressive lengthening in the return sinus node cycle with progressively premature atrial extrasystoles is not the normal behavior of the sinus node return cycle (which is rather fixed once invasion of the sinus node occurs). In these cases, marked supression of sinus node automaticity occurs by the extrasystoles and indicates severe sinus node disease.

2. *Pseudo-pseudo atrio-ventricular block.* High-degree atrio-ventricular block after a single ventricular premature beat is not normal. As previously stated, the effects of retrograde concealed conduction of a single ventricular premature beat rarely persists for more than the effects on the next P wave. This observation truly uncovers severe subnodal conduction disturbances

unmasked by the ventricular premature beat. This is not pseudo-atrio-ventricular block, but true atrio-ventricular conduction disturbances or pseudo-pseudo-atrio-ventricular block.

CONCLUSIONS

The interactions between normal and abnormal impulse formation in the human heart are extremely complex. The easiness of treatment of brady-arrhythmias by a pacemaker has decreased the interest in their study. However, from time to time, complex situations may occur simulating bradyarrhythmias which do not require pacemaker treatment. The physician should continue to make the effort to understand the exact mechanisms of a suspected bradyarrhythmia to avoid unnecessary pacemaker implantation in case of a pseudo-bradyarrhythmia.

REFERENCES

1. Langendorf R (1985) Concealed conduction: A historical review, pp 501—504 in: Zipes DP, Jalife J (eds), *Cardiac Electrophysiology and Arrhythmias.* New York: Grune & Straton Inc
2. Rosen KM, Rahimtoola SH, Gunnar RM (1970) Pseudo AV block secondary to premature non-propagated His bundle depolarizations: Documentation by His bundle electrocardiography. *Circulation* 42: 367—372

3. Cardiac pacing for bradyarrhythmias

LUC DE ROY

INTRODUCTION

One of the first descriptions of electrical stimulation of the heart was made more than 50 years ago by Hyman who delivered electrical impulses to a heart in standstill. In the mid-fifties, several publications appeared about the treatment of complete heart block with artificial external pacemakers. It was only in 1958 that Elmqvist and Senning developed the first miniature battery which was totally implanted in a patient.

Since this heroic start, the considerable technologic evolution has permitted cardiac pacing to become a daily practice therapy. From the initial indication which aimed to save lifes by preventing syncope and sudden death from profound bradyarrhythmias, pacemaker (PM) treatment has nowadays numerous other indications comprising prophylaxis and preservation of quality of life.

SICK SINUS SYNDROME

Sick sinus syndrome (SSS) has become the first indication for PM implantation in the Western countries. In Belgium, at least 1770 PM were implanted in 1989 for SSS or for related bradycardias. This represents 40% of the available PM indications. Similar data are found in other countries.

Shaw [1] in his 10-year prospective survey about the outcome of patients with SSS concluded that the overall survival did not differ from that of a normal population and that PM implantation, besides the reduction of some incapacitating symptoms, had little discernible effect on mortality. These results are somewhat disturbing and remains commonly cited in the context of treatment options of SSS. Several elements of the study of Shaw may, however, be criticized and one of the most important is the type of PM used. Although not specified, we may assume that the majority if not all of the

E. Andries, P. Brugada & R. Stroobandt (eds.), How to face 'the faces' of cardiac pacing, 29–40.

patients had VVI PM implanted. This could, of course, make a great difference as we will analyse in the next sections.

In fact, SSS is not a uniform clinical situation and we may arbitrarily divide it into (1) bradycardia, (2) pauses, or (3) bradycardia associated with tachycardia episodes. Furthermore, the ability of the sinoatrial node to react by acceleration of the heart rate, in situations of increased metabolic demand like emotional stress or physical exercise, is not invariable in this disease.

Complications in relation to pacing mode in SSS

Atrial fibrillation (AF) is a common arrhythmia in SSS. Sutton [2] analyzed the data of 21 different studies and he found in a total of 958 patients an incidence of 8.2% of AF at initial diagnosis of SSS. The incidence of AF after implantation of a PM from 19 available studies was 15.8% for a mean followup of 38 months. This figure was only 3.9% in AAI paced patients as compared to 22.3% for VVI patients. These last data are derived from a cohort of 1061 patients from 18 different studies.

Several other publications have shown similar differences in the occurrence of AF between atrial and ventricular pacing. Fülle [3] found, in a total of 223 patients, an annual incidence of AF of 3.3% in atrial pacing versus 10% in VVI pacing. At two years, the actuarial incidence was 8.7% versus 30.3%, respectively.

Rosenqvist [4] also demonstrated a comparable dissimilarity in the occurrence of permanent AF but only in the group of patients which had already paroxysmal atrial arrhythmias before PM implantation. No statistical difference was observed in the group without these previous arrhythmias.

Thus, isolated ventricular pacing seems to be more subject to the development of atrial arrhythmias in SSS. A regular atrial drive is probably one of the factors that reduces the incidence of these arrhythmias in atrial pacing. There are indeed numerous examples of arrhythmia control by regular and relative fast atrial pacing, especially in cases of bradycardia-dependent arrhythmias but also in other settings.

Data from 15 different studies reviewed by Camm [5] revealed an incidence of *thromboembolism* in SSS, for the first year after implantation by VVI pacing, of approximately 12%. Sutton [2], analyzing five different studies, and a cohort of 475 unpaced patients with SSS, found 15% thromboembolism. In one of these studies, a group of age-matched controls without SSS had only an incidence of 1.3% of systemic embolism. From five studies comparing VVI treated patients with AAI pacing, the incidence of thromboembolism was respectively, 13 and 1.6%. Similar results were observed in the reports of Santini [6] and Sasaki [7] as in several others.

Rosenqvist [4] also showed differences in the incidence of *stroke* when he compared AAI pacing versus VVI pacing but the results were not statistically significant.

The same author, however, found significant differences in *congestive heart failure* and *death rate.* Alpert [8] on the other hand, in a study of 128 patients, showed no significant difference in survival for his whole group. Nevertheless, when dual chamber pacing (DCP) was compared to ventricular pacing in patients with congestive heart failure, the difference became statistically significant in favour of DCP.

Although most of these studies are not randomized and mainly retrospective, and therefore open to criticism, there seems to be convincing arguments for a relation between VVI pacing and the incidence of AF, thromboembolism and congestive heart failure. Moreover, VVI pacing could possibly also influence survival especially in the presence of heart failure.

In addition, VVI pacing may often be complicated by a pacemaker syndrome especially in SSS patients who have frequently normal retrograde ventriculoatrial (VA) conduction.

Mechanisms of complications in SSS

The possible mechanisms for the occurrence of AF are (1) the absence of regular atrial drive, (2) the presence of retrograde conduction, and (3) atrial enlargement favoured by VVI pacing. Thromboembolism could be explained by the higher incidence of AF in VVI patients but AF was not always strictly related with this complication, as in the study of Curzi [9]. One-to-one retrograde VA conduction has been evoked by this author as the major factor for the occurrence of thromboembolism. In patients without 1/1 VA conduction, thromboembolism seemed to be rare and related to the onset of AF.

Congestive heart failure has been attributed to the loss of atrioventricular (AV) synchronism leading to an excess of catecholamine release, responsible for abnormal left ventricular wall motion.

Retrograde VA conduction could on the other hand increase the production of atrial natriuretic factor an venous regurgitation, leading to a decrease of rapid diastolic filling and stroke volume. The higher incidence of chronic AF in VVI paced patients could also be incriminated as a cause of CHF but VVI pacing in itself seems to play an important role [4].

Considering these different complications in relation to VVI pacing, we could hypothesize that the absence of any beneficial effect on survival in paced SSS patients is possibly related to a worse outlook with this mode of pacing than with atrial pacing. Thus, AAI or DCP could maybe favourably influence the natural history of SSS with respect to morbidity and mortality. In addition, one might speculate about the benefit of prophylactic PM insertion in overt sinus node disease even without significant symptoms.

Choice of the ideal pacing mode in SSS

VVI pacing is no more acceptable for the treatment of the SSS if we consider the above-mentioned complications.

Atrial pacing is obviously an important factor in the management of sinoatrial disorders. If we analyze the advantages and the disadvantages of the different atrial pacing modes, AAI or DCP, we may notice that *AV synchronism* is maintained and that *atrial arrhythmias* can be controlled by the two pacing modes, in contrast with VVI pacing. *Rate adaptability* can be achieved by any pacing mode if Rate Response PM (VVIR, DDDR, DDIR) are used or if chronotropic competence is still present — as it is the case in 50 to 60% of SSS patients — when AAI or DCP are utilized.

The disadvantages of DDD PM are the possible occurrence of *PM mediated tachycardias* or *rapid ventricular response* following supraventricular arrhythmias. Different algorithms are now available to avoid these problems.

The *morbidity* due to more complex implantation procedures is somewhat higher in DCP than in AAI PM. In contrast, morbidity and possibly mortality remain some of the major drawbacks in VVI pacing, as mentioned before. The *cost* is obviously higher in DCP than in single chamber pacing.

One of the problems of AAI pacing is the occurrence of *AV conduction disturbances.* Sutton [2] found an incidence of 16.6% of AV conduction abnormalities at initial diagnosis of SSS and over a mean follow-up period of 34 months an occurrence of 8.4%. The definition of these AV abnormalities was, however, relatively large: first-degree AV block, bundle branch block and AV node Wenckebach point equal or lower than 120/min. In the study of Rosenqvist [4], the incidence of second degree AV block was of 4.5% at 4 years. This represents a yearly incidence of 1.1%. Similar numbers are reported by several other authors.

Despite these relatively low percentages of significant conduction disturbances, we have to consider the possible evolution of latent AV conduction disturbances over the years, taking into account the longevity of modern PM. This is all the more so since the use of antiarrhythmic drugs may be required in these patients in order to maintain a stable sinus rhythm. These and other medications may alter AV conduction disturbances which were of no clinical significance before.

Another question which remains unsolved is the negative predictive value of an *AV Wenckebach point* above 120/min in patients without any evidence of conduction disturbance prior to PM implantation. This issue was prospectively analyzed by the group of Camm [10] in a small series of 24 patients. They found a 17% rate of progression of AV block requiring revision of PM system within 6 months. These results are quite different from those reported in previous retrospective studies. One of the most striking elements of their findings was that in 3 of the 4 patients with the occurrence of AV block, the AV Wenckebach point was above 120/min. This raises the question about the validity of the cutoff point of 120/min and about the necessity of measuring the HV interval during incremental atrial pacing. This would, however, also suggest that every patient being paced for sinoatrial disease should undergo an electrophysiological study, which constitutes an unacceptable burden for the PM centres. Systematic implantation of a DDD PM represents, on the other hand, a considerable financial charge and this question has nowadays no fully acceptable solution.

About half of the patients with SSS will have inadequate rate adaptation at exercise, having an impaired chronotropic response with lowered heart rate reserve. This is called *chronotropic incompetence* which may limit the effort tolerance and, consequently, the quality of life of some patients. Until some years ago, only single-chamber PM had rate adaptation capabilities and exclusively AAIR or VVIR devices could be used in these circumstances. At present, several types of DDDR or DDIR systems are readily available and commonly inserted for chronotropic incompetence. As demonstrated by Wilkoff [11], cardiac chronotropic response is a simple linear mathematical function of exercise intensity, age, resting heart rate and maximal function capacity and is independent of the exercise protocol, allowing easy clinical evaluation.

From a practical point of view, we may say that SSS remains the most frequent indication for PM implantation in our countries. Survival of patients suffering from this disease seemed not to be influenced by this treatment to date. Atrial or DCP could possibly improve the prognosis of these patients if we consider the lower incidence of complications as compared to VVI pacing. Single ventricular pacing should therefore be abandoned in SSS.

When AV conduction disturbances are present or suspected, we would choose a DDD PM (Table 1). When AV conduction is unimpaired, AAI PM, or DDI mode will be favoured. As we saw, assessment of the AV conduction is not always a simple task. We may nevertheless assume that AV conduction is and will probably remain normal if no ECG conduction abnormalities are present, if AV Wenckebach point is far above the theoretical limit of 120/min, if no significant atrial arrhythmias are documented and if no negative dromotropic treatment is expected. All these conditions and the physician's fear of incomplete protection when no ventricular back-up is available, will obviously limit the insertion of AAI PM in practice, as we notice from the data of the PM register (Table 2). Indeed, only 0.5% of all the PM implanted in Belgium in 1989 were of the AAI type, while SSS or bradycardia accounted for more than 40% of all the PM implantations. Moreover, this number has remained stable since 1982, despite better definitions about PM indications and ideal type of PM to be used.

Table 1. Choice of pacemaker type in SSS.

Chronotropic competence	— no AV conduction disturbances	→ AAI (DDI)
	— possible AV conduction disturbances	→ DDD
Chronotropic incompetence	— no AV conduction disturbances	→ AAIR (DDIR)
	— possible AV conduction disturbances	→ DDDR
Atrial arrhythmias (rare)		→ DCP + overdrive capabilities (AAI?)
Atrial arrhythmias (frequent)		→ DCP + overdrive capabilities

Table 2. Data from the Belgium Pacemaker Registry, 1989.

Type of pacemaker	Total number of patents	Percentage
VVI	3028	49.8
VVIR	990	16.3
DDR	1780	29.3
DDDR	243	4.0
AAI	30	0.5
AAIR	10	0.16
Total	6017	100

CAROTID SINUS SYNDROME

The carotid sinus syndrome (CSS) is characterized by intermittent excessive vagal tone originating from the hypersensitive carotid sinus which produces undue slowing of the heart rate and/or the AV conduction. It is the cardio-inhibitory reflex. The CSS can be complicated by an inappropriate blood pressure drop, the so-called vasodepressive component. Finally, there are mixed types which combine bradycardia and hypotension. Graux [12] found a 26% of CSS in a population of 245 patients with syncope or dizziness from unknown origin. Among them, 27% had a cardioinhibitory response, 22% a vasodepressive and 51% a mixed type.

In CSS, AV conduction delay is often hidden by the excessive slowing of the sinoatrial node. The effect on the AV node can be unmasked by atrial pacing during carotid sinus massage. The conjunction of AV conduction disturbances constitutes a definite indication for ventricular pacing by VVI or DCP. However, as mentioned before, vasodepressor problems complicates the syndrome in almost 75% of the cases and this may cause major problems in the future management of these patients. That's the reason why DCP, often with a rather fast rate, remains the treatment of choice in this syndrome. Nevertheless, DCP is frequently associated with inopportune ventricular pacing with asymmetric ventricular depolarization or competition and unnecessary drainage of battery energy when normal AV conduction is preserved.

New algorithms are presently proposed by the manufacturers in order to face these problems and those of persisting symptoms despite PM implantation. Some of them provides the possibility of regular AAI pacing with an automatic switch to DDD mode when AV conduction block appears [13]. An automatic atrial rate increase in case of intervening bradycardia is another characteristic of some of these algorithms.

In conclusion (Table 3), one can state that AAI pacing may be proposed if AV conduction disturbances have been ruled out by an extensive evaluation

Table 3. Choice of pacemaker type in the carotid sinus syndrome.

Condition		Type of pacemaker
No AV conduction disturbances	→	AAI?
AV conduction disturbances		
— no vasodepressive reaction		
— no pacemaker syndrome	→	VVI?
Other situations	→	DCP

comprizing atrial pacing. VVI pacing should only be indicated if there is no significant blood pressure fall with this kind of pacing compared to DDD pacing and in the absence of PM syndrome. Vasodepressor reaction and AV conduction slowing are, however, variable and reliable exclusion remains questionable. VVI pacing should therefore only be exceptionally, if ever, used in the setting of CSS. DCP is indicated in all the other instances and the newer developments will certainly increase the effectiveness of this pacing mode in the treatment of the remaining symptoms of these patients.

ATRIOVENTRICULAR BLOCK

Considering that survival in isolated first-degree AV block is not influenced by PM implantation, one may identify only second- and third-degree block as possible indications for PM therapy. They may be permanent or intermittent. *Acquired permanent or intermittent third-degree AV block* with symptomatic bradycardia or congestive heart failure, is clearly an indication for PM therapy [14]. The same applies for *second-degree AV block* and *atrial fibrillation* with complete, advanced AV block or pronounced bradycardia, unrelated to digitalis or other drugs known to impair AV conduction. However, a clear relationship between symptoms and the conduction disturbance has to be presumed. In the absence of definite symptoms, the decision for PM implantation becomes more difficult.

Asymptomatic-acquired third-degree AV block seems to be an indication because of the irremediable evolution towards a progressive slowing of the escape rhythms.

Congenital third-degree AV block, however, may remain asymptomatic for many years. Dewey [15] identified several criteria indicating the presence of an unstable junctional escape rhythm, requiring prophylactic PM implantation. He followed, prospectively, 27 patients with congenital complete heart block for a mean period of 8 ± 3 years with frequent ambulatory Holter monitoring. The conclusion of his study was that a mean day-time junctional rate below 50 beats per minute associated with other evidence of an unstable junctional escape mechanism is predictive of cardiac complications as sudden death, syncope, presyncope or excessive fatigue. This category of patients should probably undergo prophylactic PM implantation. On the other hand, patients with a mean day-time heart rate of 50 bpm or more didn't have any adverse clinical outcome.

Asymptomatic second-degree AV block type II is generally considered as a definite prophylactic indication because of the unpredictable and often severe outcome of this usually infranodal conduction defect.

Type I second-degree AV block has generally been assumed to be benign, especially in the absence of organic heart disease [16]. Young [17], however, suggested that this kind of block probably represents a significant disease of the cardiac conduction system which may be progressive with a guarded prognosis. This statement was based on the occurrence of 44% of complete heart block in his population of children and adolescents during a followup of 1 to 18 years. Shaw [18] examined the outcome of 214 patients with chronic second-degree AV block during a 14-year prospective study. His findings were that patients with chronic Mobitz type I block have a similar unfavourable prognosis to that of Mobitz type II block and that those patients fitted with PM had a 5-year survival of 78%, which was similar to that expected for the normal population. On the contrary, the unpaced patients did very badly with a 5-year survival of 41%, regardless the type of block, type I or type II. These results seem to refute the benign reputation of chronic Mobitz type I block and imply that this kind of block should be considered for PM implantation. Such an attitude infers, however, a substantial increase of prophylactic PM insertions with the inherent cost increment [19].

Type of pacemaker for atrioventricular block

In chronic second-degree or third-degree AV block, AV conduction is not able to recover and if PM insertion is indicated, at least ventricular pacing has to be ensured. The fundamental question however persists: Is only ventricular pacing required and is atrial synchronization useful?

VVI versus atrial triggered PM. When these two pacing modes were compared by Kruse and Ryden [20], they not only observed a lower underlying atrial rate in VDD pacing but also a significantly higher achieved work load, indicating an unquestionable benefit for atrial synchronism. If one might opt for VVI pacing, one has first of all to rule out the presence of *retrograde conduction* (Table 4). The frequency of VA retrograde conduction is uncertain, depending on the investigating method used, the level of negative dromotropic drug impregnation and the sensitivity of atrial capture to the ventricular pacing. Moreover, Cazeau et al. [21] analyzed the behaviour of the VA conduction at rest and during exercise in 17 patients. They found that 6 of the 10 patients without VA conduction at rest recover retrograde conduction at exercise and that this phenomenon did not obey any general rule nor could be predicted individually. It thus seems that VVI pacing could

Table 4. Choice of pacemaker type in atrioventricular block.

Condition			Type of pacemaker
Retrograde conduction:	— present	→	DCP
	— absent: at rest + exercise	→	VVIR?
	at rest only	→	DCP
Pacemaker syndrome:	unpredictable!	→	DCP
Chronotropic response:	— normal	→	DDD
	— abnormal	→	DDDR

lead to unexpected side effects caused by unsuspected recovery of retrograde conduction.

Furthermore, even in the absence of retrograde conduction, *pacemaker syndrome* may occur. The incidence of this syndrome is probably underestimated, evaluated at 5—15% according to the literature. However, some recent studies showed a significantly higher incidence. Pacemaker syndrome indeed occurs in the majority of patients who served as their own control when VVI and DCP were compared. In fact, often no objectively measured difference could be found to corroborate this observation but the evidence was that a great proportion of the patients studied expressed a clear preference for dual chamber pacing. Heldman [22], in a blind, randomized trial of 40 unselected patients, found an incidence of 83% of clinical pacemaker syndrome when VVI pacing was programmed with 65% of patients experiencing moderate to severe symptoms. There were, however, no readily identified clinical, hemodynamic, or electrophysiological parameters that predicted which patients would develop pacemaker syndrome. The fact is that most of the patients are not aware of milder symptoms related to VVI pacing when there is no basis for comparison and the true incidence of pacemaker syndrome remains difficult to assess in a population of patients with a single-chamber pacing system.

Rate response versus fixed rate pacing

VVI versus VVIR pacemakers. The benefit of rate response (RR) of single chamber PM has been extensively demonstrated with a less perceived exertion level, a lower respiratory rate, a higher cardiac index and output, and even a reduction of rest stroke-volume and left ventricular size over time. Stroke-volume index during exercise is, however, higher in VVI pacing, indicating that adaptation mechanism compensate to some extent.

VVIR versus atrial triggered pacemakers (Table 5). It has also largely been demonstrated that the major factor for exercise tolerance remains the rate adaptability rather than AV synchronism, as proven by an identical exercise capacity when comparing VVIR and VAT pacing. However, some studies

Table 5. Choice between DDD *or* rate response pacemakers.*

	VVIR	DDD	DDDR
Atrial transport	–	+	+
Atrial synchrony	–	+	+
Retrograde conduction	+/–	PMT	PMT
Rate adaptability	+	+/–	+

* Indication of presence (+) or absence (–) of the item in each category of pacing mode. PMT signifies that pacemaker mediated tachycardias may occur, but that adequate prevention or interruption capabilities are available. See text for discussion.

have shown an additional benefit for atrial triggering during exercise, leading to a further increase of cardiac output when compared to rate-response.

The advantage of dual-chamber pacing has, on the other hand, probably been underestimated because insufficient attention has been given to appropriate AV delay programming, a factor which has a major importance in left ventricular filling. Wish and co-worker [23] underscored the essential role in adequately programming the AV delay with respect to the timing between right atrial stimulation and left atrial contraction, the latter being the most important determinant of end-diastolic volume and stroke-volume of the left ventricle. Noninvasive evaluation of an appropriate timing of the left atrial contraction can be achieved by Doppler echocardiography of the mitral flow. The optimal situation is when the mitral 'A' wave arrives simultaneously with the ventricular pacing spike.

In a study of 14 patients with high-degree AV block, without heart failure, Menozzi [24] compared symptoms and hemodynamic parameters in chronic VVIR and DDD pacing in the same patients in order to assess the benefits of both pacing modes. The symptom score between the two groups was statistically different and largely in favour of DDD pacing, with intolerable symptoms in 36% of the patients in VVIR mode, who needed early switching to DDD pacing. Cardiac output and atrial natriuretic peptide were also significantly better in DDD pacing, despite a similar effort tolerance. The conclusion of the author is that atrial synchronization reflects a better quality of life for paced patients, even if individual variability exists. On the other hand, arterial pressure is not always an adequate indicator of the hemodynamic benefit, the major advantage being the fall in venous pressure.

In addition, as stated by Levine [25], despite the fact that VVIR pacing may result in better exercise capabilities in patients with impaired chronotropic competence, DCP has major advantages over ventricular pacing at rest and during mild activity.

Furthermore, rate response VVI pacing may compete with sinus rhythm at some exercise levels and cause blood pressure fall when AV synchronism is lost or retrograde conduction appears.

Dual chamber rate response pacemakers. If however chronotropic response is reduced or absent, the new generation of DDDR PM may confer an adequate way to provide rate adaptation at exercise. Moreover, if the heart rate reserve is maintained, there is no benefit of rate-response ventricular pacing, the sinus node being the most physiological parameter to track ventricular pacing.

From the aforementioned data, it appears that DCP remains the most physiological kind of pacing in AV block. Single chamber pacing with VVI PM should be used with caution because of the possible and unexpected occurrence of retrograde conduction and the likelihood of some degree of pacemaker syndrome which may alter the quality of life of the patients. The primary goal of pacemaker therapy remains, in addition of the control of bradycardia, to restore the heart function as close as possible to normal by providing AV synchronism and rate adaptability.

REFERENCES

1. Shaw DB, Holman RR, Gowers JI (1980) Survival in sinoatrial disorder (sick sinus syndrome). *Br Med J* 21: 139—141
2. Sutton R, Kenny RA (1986) VVI versus AAI or dual chamber pacing in Sick sinus syndrome, pp 253—265 in: Santini M, Pistolese M, Alliegro A (eds), *Progress in Clinical Pacing.* Rome
3. Fülle P, Markewitz A, Weinhold C (1986) How to pace a sick sinus syndrome, pp 330—342 in: Santini M, Pistolese M, Alliegro A (eds), *Progress in Clinical Pacing.* Rome
4. Rosenqvist M, Brandt J, Schüller H (1988) Long-term pacing in sinus node disease: Effects of stimulation mode on cardiovascular morbidity and mortality. *Am Heart J* 116: 16—22
5. Camm AJ, Katritsis D (1990) Ventricular pacing for sick sinus syndrome: A risky business? *PACE* 13: 695—699
6. Santini M, Alexidou G, Ansalone G, Cacciatore G, Cini R, Turitto G (1990) Relation of prognosis in sick sinus syndrome to age, conduction defects and modes of permanent cardiac pacing. *Am J Cardiol* 65: 729—735
7. Sasaki S, Shimotori M, Akahane K, Yonekura H, Hirano K, Endoh R, Koike S, Kawa S, Furuta S, Homma T (1988) Long term follow-up of patients with sick sinus syndrome: A comparison of clinical aspects among unpaced, ventricular inhibited paced and physiologically paced groups. *PACE* 11: 1575—1583
8. Alpert MA, Curtis JJ, Sanfelippo JF, Flaker GC, Walls JT, Mukerji V, Villarreal D, Katti SK, Madigan NP, Morgan RJ (1987) Comparative survival following permanent ventricular and dual-chamber pacing for patients with chronic symptomatic sinus node dysfunction with and without congestive heart failure. *Am Heart J* 113: 958—965
9. Curzi GF, Mocchegiani R, Ciampani N, Pasetti L, Berrettini U, Purcaro A (1985) Thromboembolism during VVI permanent pacing, pp 1203-1206 in: Gomez FP (ed), *Cardiac Pacing.* Madrid: Editorial Grouz
10. Haywood GA, Ward J, Camm AJ (1990) Atrioventricular Wenckebach point and progression to atrioventricular block in sinoatrial disease. *PACE* 13: 2054—2058
11. Wilkoff BL, Corey J, Blackburn G (1989) A mathematical model of the cardiac chronotropic response to exercise. *J Electrophysiol* 3: 176—180
12. Graux P, Merkerke W, Lemaire N, Beaugeard D, Cornaert P, Dubeaux P-A, Dutoit A, Croccel L (1988) Le syndrome du sinus carotidien: Etude hémodynamique prospective

de 245 patients: Incidence de ses formes cliniques et propositions de stratégie thérapeutique. *Stimucœur* 3: 183–189
13. Girodo S, Ritter P, Pioger G, Lamaison D, Malherbe O (1990) Improved dual chamber pacing mode in paroxysmal atrioventricular conduction disorders. *PACE* 13: 2059–2064
14. Frye RL, Collins JJ, DeSanctis RW, Dodge HT, Dreifus LS, Fish C et al. (1984) Guidelines for permanent cardiac pacemaker implantation. *Circulation* 70: 331A–339A
15. Dewey RC, Capeless MA, Levy AM (1987) Use of ambulatory electrocardiographic monitoring to identify high-risk patients with congenital complete heart block. *N Engl J Med* 316: 835–839
16. Strasberg B, Amat-Y-Leon F, Dhingra RC, Palileo E, Swiryn S, Bauernfeind R, Wyndham C, Rosen KM (1981) Natural history of chronic second-degree atrioventricular nodal block. *Circulation* 63: 1043–1049
17. Young D, Eisenberg R, Fish B, Fisher JD (1977) Wenchebach atrioventricular block (Mobitz type I) in children and adolescents. *Am J Cardiol* 40: 393–399
18. Shaw DB, Kekwick CA, Veale D, Gowers J, Whistance T (1985) Survival in second degree atrioventricular block. *Br Heart J* 53: 587–593
19. Campbell RW (1985) Chronic Mobitz type I second degree atrioventricular block: Has its importance been underestimated? *Br Heart J* 53: 585–586
20. Kruse I, Ryden L (1981) Comparison of physical work capacity and systolic time intervals with ventricular inhibited and atrial synchronous ventricular inhibited pacing. *Br Heart J* 46: 129
21. Cazeau S, Daubert C, Mabo P, Ritter P, Lelong B, Pouillot C, Paillard F (1990) Dynamic electrophysiology of ventriculoatrial conduction: Implications for DDD and DDDR pacing. *PACE* 13: 1646–1655
22. Heldmann D, Mulvihill D, Nguyen H, Messenger JC, Rylaarsdam A, Evans K, Castellanet MJ (1990) True incidence of pacemaker syndrome. *PACE* 13: 1742–1750
23. Wish M, Fletcher RD, Cohen A (1989) Hemodynamics of AV synchrony and rate. *J Electrophysiol* 3: 170–175
24. Menozzi C, Brignole M, Moracchini PV, Lolli G, Bacchi M, Tesorieri MC, Tosoni GD, Bolloni R (1990) Intrapatient comparison between chronic VVIR and DDD pacing in patients affected by high degree AV block without heart failure. *PACE* 13: 1816–1822
25. Levine P, Mace R (1983) The pacemaker syndrome, p 3 in: *Pacing therapy: A Guide to Cardiac Pacing for Optimum Hemodynamic Benefit.* Mount Kisco, New York: Futura Publishing Comp

4. From VVI to DDD pacemakers: Glossary of terms and normal functions

ALFONS SINNAEVE

IDENTIFICATION CODE (OR WHAT'S IN A NAME?)

The three-letter code is still used throughout the world to denote pacing mode. This code was established by the Inter-Society Commission for Heart Disease Resources (ICHD) as a standardized means for identifying the functional operation of a pulse generator, irrespective of model number, trade name or complexity. In this initial ICHD code, the first letter indicates the paced chamber(s) and the second letter the chamber(s) being sensed while the third letter represents the mode of response to a sensed event, i.e. how the pacemaker will react to cardiac electrical activity (Figure 1).

Some examples may help to understand the meaning of these three letters:

- VVI: this is a pacemaker which is only pacing and sensing in the ventricle and works in the inhibited mode, i.e. a ventricular sense inhibits the ventricular pacing.
- VAT: stands for a pacemaker able to pace in the ventricle when an electrical event is sensed in the atrium, i.e. an atrial event triggers a ventricular stimulus.
- DVI: means a pacemaker which will pace in both chambers but senses only in the ventricle and works only in the inhibited mode; in the absence of ventricular activity, sequential stimuli will be delivered to both the atrium and the ventricle.
- VDD: indicates a pacemaker which is only pacing in the ventricle but has a dual-chamber sensing ability and a double mode of response, i.e. an atrial sense triggers a ventricular stimulus and a ventricular sense inhibits the ventricular stimulus.
- DDD: denotes a pacemaker which is pacing and sensing in both atrium and ventricle and works just as well in the triggered as in the inhibited mode, e.g. an atrial spontaneous event can inhibit the atrial output and might just as well trigger a ventricular stimulus.

With the development of more complex pacemakers, the three-position code

E. Andries, P. Brugada & R. Stroobandt (eds.), How to face 'the faces' of cardiac pacing, 41–81.

- **LETTER 1 : CHAMBER PACED**
  - A = atrium
  - V = ventricle
  - D = dual (atrium and ventricle)
- **LETTER 2 : CHAMBER SENSED**
  - A = atrium
  - V = ventricle
  - D = dual (atrium and ventricle)
  - O = none
- **LETTER 3 : MODE of RESPONSE**
  - I = inhibits pacing
  - T = triggers pacing
  - D = dual (triggered and/or inhibited)
  - O = none
  - R = reverse

Figure 1. Three-letter identification code.

was expanded to the five-position code: the fourth letter stands for the programmability features and the fifth letter indicates the special tachyarrhythmia functions. Nevertheless, the three-position code seems to be sufficient for most purposes and still facilitates communication among physicians and others in the field of pacing.

The *rate responsive pacing mode* (also called rate-adaptive or rate-modulated) was not known yet when the three-letter code was set up, so the additional 'R' in the fourth position stands for rate responsiveness, e.g.:

VVIR: is a ventricular pacing and sensing pulse generator working in the inhibited mode, but capable of adapting the rate to the metabolic demands of the patient by means of a sensor.

DDDR: is similar to DDD pacing but because of the presence of an additional sensor, it can provide rate-responsive pacing in the absence of an appropriate sinus node response.

ESSENTIAL PARTS OF A VVI PACEMAKER

For a proper understanding of the more complex DDD pacemaker, it is necessary to be familiar with the basic elements of the simple VVI pulse generator.

The VVI pulse generator is *R wave inhibited*, i.e. the pacemaker impulses are blocked by the ventricular depolarizations if the natural heart rate is higher than the programmed pacemaker frequency. Since the pulse generator only delivers impulses to the heart when the spontaneous heart rate is too low (or if the pauses are too long), this pacemaker is working in a *on-demand mode* or *stand-by mode.*

The advantages of the on-demand pacemaker with regard to the asynchronous-type working at a fixed rate are twofold:

* on-demand types avoid detrimental competitive rhythms (stimulus on the T wave),
* on-demand types enhance pacemaker longevity by decreasing the mean current drain from the battery (saving energy).

A VVI pulse generator consists of three essential modules: a timer, an output circuit, and a sensing circuit (Figure 2).

The *timer* is the heart of the pacemaker. In recently developed devices, the timing element is made up of an *oscillator* and some *counters.* Because they deliver a very stable frequency, modern oscillators are built around a piezoelectric quartz crystal such as used in a digital watch. The frequency at which the crystal vibrates depends on its physical characteristics and is adjusted to 32 kHz. A specified number of crystal oscillations establishes a pacing interval. A digital counter adds up the appropriate number of impulses delivered by the oscillator with an accuracy of one oscillation. This extremely accurate timing base allows noninvasive reprogramming of the pulse generator with very small time increments, e.g. tenths of a millisecond for a pulse width.

The *output circuit* determines the shape (constant-current or voltage type) and the amplitude of the pacing impulses delivered to the heart. It contains the *output capacitor* which is charging from the battery with a small current via a large resistor and which is discharging with a large current through the heart tissue and via a closed switch to stimulate the heart. In the pacemaker, the switch is replaced by a transistor and is controlled by the timing circuit.

Voltage doublers are often used in the output circuit to give a higher

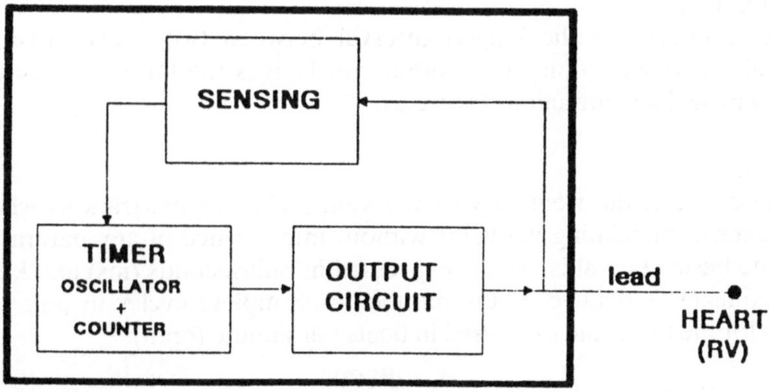

Figure 2. Essential parts of a VVI pacemaker (R wave inhibited; on demand; stand-by).

voltage for stimulation, while the rest of the circuitry operates at a lower voltage. In this type of circuit, the output capacitor is in parallel with the battery during the charging period and is switched in series position during the quick discharging period. The main advantage of this concept is that no extra batteries are needed for an adequate stimulation safety margin, so the pulse generator can be smaller and lighter.

Finally it should be noted that all output circuits have to be protected by a *Zener-diode* against excessive high voltages which can be generated during defibrillation.

The *sensing circuit* monitors the heart's activity. This circuit can be thought of as a receiver that both detects heart activity (normal contractions and PVC's) and discriminates against other signals to make the pacing circuitry react appropriately. Unfortunately, the depolarization waves of the heart which are to be sensed, vary greatly from patient to patient in frequency, amplitude, slew rate, duration, and direction. For that reason, the sensing circuit is a rather complex building block consisting of different parts:

* A *differential amplifier* receives the weak signals from the lead and amplifies them to facilitate further processing.
* A *bandpass filter* discriminates by frequency in an attempt to reject the T waves (which have lower frequency components) and myopotentials as well as EMI signals (which both have higher frequency components).
* A *level detector* discriminates by amplitude (+ and −) and diminishes the risk for false inhibition by P wave sensing in the ventricle.
* A *refractory period generator* ensures that the pacemaker does not sense its own resultant depolarization or the corresponding T wave.

BASIC CONCEPTS AND DEFINITIONS OF VVI PACING

Basic interval:

The basic interval is the longest interval between two consecutive paced ventricular depolarizations or, in other words, it is the longest pause after a paced ventricular contraction (Figure 3).

Basic rate:

The basic rate is the frequency of the ventricular depolarizations when the pacemaker is stimulating the heart without interference of any natural beats. Since the basic interval is usually expressed in milliseconds (ms) and knowing that frequency is defined as the number of complete cycles in unit time, it follows for the basic rate expressed in beats per minute (bpm):

$$\text{Basic Rate (bpm)} = \frac{60\,000}{\text{Basic Interval (ms)}}$$

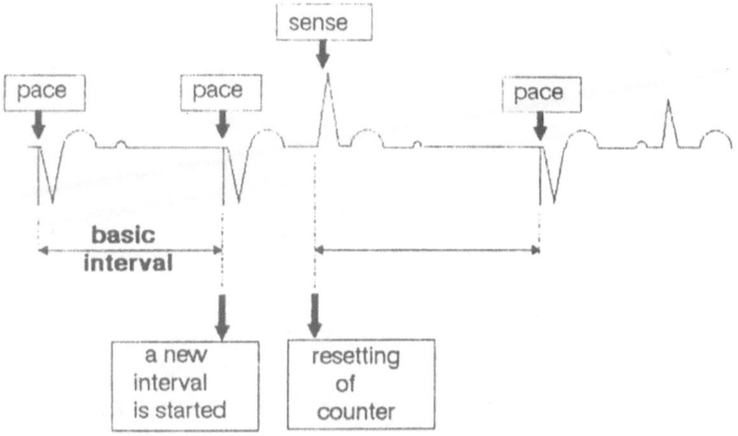

Figure 3. Definition of basic interval.

and also

$$\text{Basic Interval (ms)} = \frac{60\,000}{\text{Basic Rate (bpm)}}$$

Escape interval:
The escape interval is the longest interval between a sensed spontaneous ventricular event and the following paced event or it is the longest pause after a natural occurring ventricular contraction (PVC or normal QRS) (Figure 4).

Escape rate:
The escape rate is the lowest ventricular frequency to which the heart rate can decrease before the pacemaker takes over and starts pacing. The escape rate is the inverse of the escape interval, so it follows in the same way as for the basic rate:

$$\text{Escape Rate (bpm)} = \frac{60\,000}{\text{Escape Interval (ms)}}$$

Hysteresis:
Hysteresis is a delay in the change from spontaneous heart rate to ventricular paced rate, allowing the heart frequency to decrease below the programmed basic rate.

The different responses of pulse generators with the hysteresis function

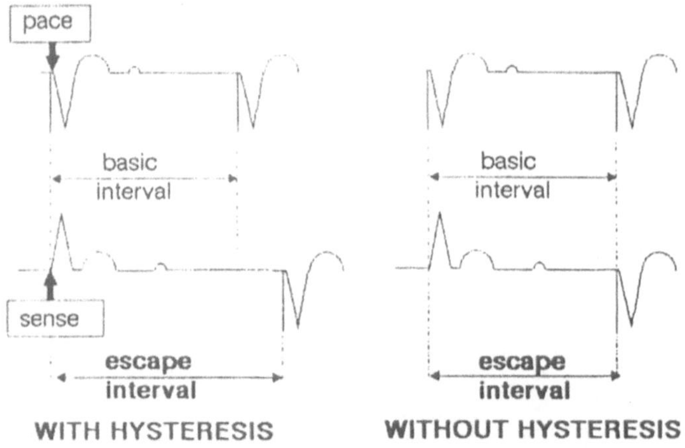

Figure 4. Definition of escape interval.

switched on and off are easily deduced from Figure 5. The diagram at the top shows the fluctuation of the spontaneous heart rhythm as a function of time. Without hysteresis, the pulse generator will start stimulation at basic rate as soon as the intrinsic ventricular rhythm drops below the rate. On the contrary, when the hysteresis is activated, the device does not start pacing until the ventricular rhythm becomes lower than the escape rate.

Obviously, hysteresis means that the escape interval exceeds the basic interval. However, the *escape interval equals the basic interval in the absence of hysteresis* (or if the hysteresis mode is switched off).

Although not all pacemaker types have the hysteresis possibility and some physicians are even pessimistic about that specific programmable function, it might have several advantages:

* hysteresis gives a better preservation of sinus rhythm,
* for heart rates between the basic rate and the escape rate, it causes a modest increase in cardiac output due to the atrial kick (especially important in patients with a poor ventricular function),
* to some extent it avoids competition between paced and spontaneous rhythm,
* hysteresis increases also the ability to obtain a diagnostic electrocardiogram,
* last but not least it prolongs battery life since sensing uses a lot less power than pacing!

Nevertheless, with hysteresis there are also some disadvantages:

* confusion in the interpretation of the electrocardiogram is possible if the physicians who participate in the care of cardiac patients are not

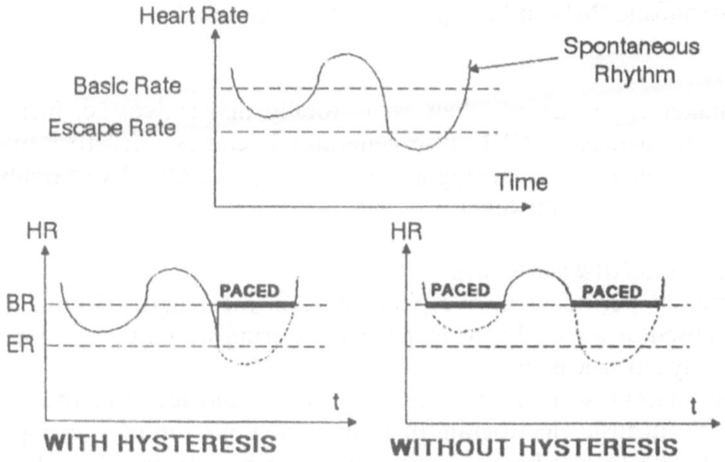

Figure 5. Variation of heart rate and paced rate with and without hysteresis.

familiar with hysteresis: *prolonged intervals* are expected with this mode of pacing and *do not indicate dysfunction.*

* if the basic pacing rate at rest is replaced by a slower intrinsic rate during activity a *paradoxical slowing of the ventricular rate* can occur (the slow natural rhythm which is mostly not of sinus origin and occurs during exercise, must have a rate between the basic and the escape rates).

Sensitivity:

Sensitivity is sometimes referred as the *sensing threshold.* Sensitivity of a pacemaker means the ability of the sensing circuit to respond to weak input signals. It is the required minimum input signal (expressed in millivolt) that will reset the timer and inhibit the output circuit of the pulse generator.

A device is more sensitive when it reacts on a smaller input signal, so it should be noted that:

* *high sensitivity* corresponds to a *small number* of millivolt (mV),
* *low sensitivity* is expressed as a *large number* of millivolt.

The numerical value of pacemaker sensitivity may be expressed in terms of an arbitrary signal using a square wave, a triangular waveform, a haversine ($\sin^2$), etc. Unfortunately, for the time being, there is no universal standard for test signals, making it very difficult or impossible to compare the published specifications of different manufacturers. Moreover, none of the many test signals bears much resemblance to an intracardiac signal, either in the time or in the frequency domain. Consequently, a strict control of the sensing capability at the implantation and during follow-up is strongly recommended (cf. avoiding undersensing and oversensing: Chapter 5 by Prof. Roland Stroobandt MD and Chapter 8 by Marc Goethals).

Refractory:
A pacemaker is refractory when he is totally insensitive, i.e. incapable of detecting any signals. A VVI pulse generator becomes refractory by switching off the sensing circuit making it impossible for a natural ventricular event (QRS or PVC) to reset the timer.

Refractory period of a pacemaker:
The refractory period of a VVI pacemaker is the length of time following either a sensed or a paced ventricular event during which the pacemaker will not sense any intrinsic activity.

In some pacemakers, the refractory period is shorter after the sensing of the patient's intrinsic electrogram than after the emission of a pacing impulse (Figure 6). The logical basis of this difference is that paced beats are normally followed by wider QRS complexes.

For the proper functioning of an on-demand pacemaker, the refractory period is a crucial requirement because it prevents the undesirable sensing of:

* *the pulse generators own output impulse* (the output spike is prevented from re-entering in the sensing system in order to avoid saturation of the very sensitive differential amplifier),
* *the QRS complex elicited by the PM stimulus,*
* *the T wave following a paced beat* (theoretically, the filter can reject most of the T waves but with the occurrence of acute ischemia or electrolyte disturbances, the T wave may become more pronounced and may confuse the sensing circuit),
* *afterpotentials generated at the electrode-tissue interface* (due to polarization phenomena at this interface, a charge and consequently a voltage

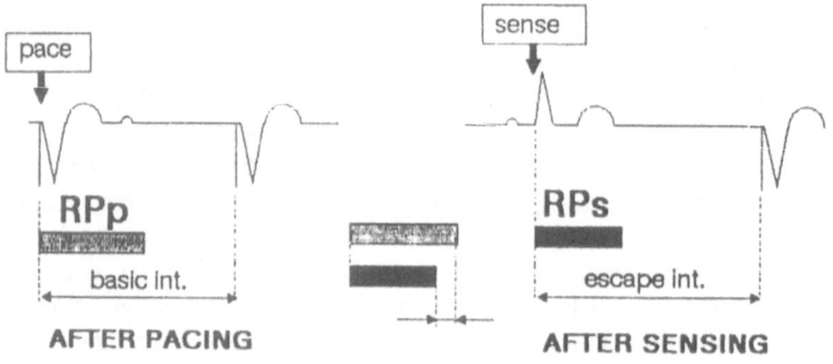

Figure 6. Refractoriness of VVI pacemakers. RPp = paced refractory period, RPs = sensed refractory period.

is built up which disappears slowly while the output capacitor is recharging after the delivery of a stimulating impulse; especially with small electrode surfaces, large impulse amplitudes and long pulse durations, this afterpotentials may become important and, in combination with the remainder of the T wave, they could be misinterpreted by the sensing circuit and inadvertently reset the counter).

In most of the recently developed pacemakers, the refractory period is programmable and has to be adjusted for an optimal duration:

* a refractory period that is *too short* may allow sensing of the T wave or even the late portion of the QRS complex (over-sensing),
* with a refractory period which is *too long*, the sensing circuit will fail to detect some spontaneous beats and the pulse generator may act in a fixed-rate mode (under-sensing), creating the possibility of delivering a spike in the ventricular vulnerable period with the potential danger of precipitating ventricular fibrillation.

Some pacemakers have a total refractory period consisting of two consecutive parts: an *absolute refractory period* followed by a *relative refractory period*. During the absolute refractory period, the sensing circuit is completely unable to detect any signals. On the contrary, during the relative refractory period, the pacemaker is able to sense. However, upon detection of a signal during this period, the pacemaker will not react with resetting of the counter and inhibition of the output circuit, but will start a new relative refractory period. If the relative refractory period is rather short (so-called *interrogation increments* or *noise sampling periods* of, e.g., 60 ms) it can be used to identify electromagnetic interference (EMI). Signals occurring with a rather high frequency (higher than 17 Hz for a noise sampling period of 60 ms) are detected during every interrogation increment and will cause the

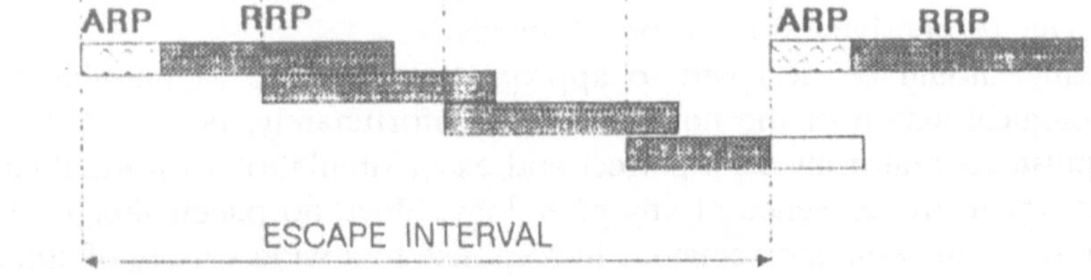

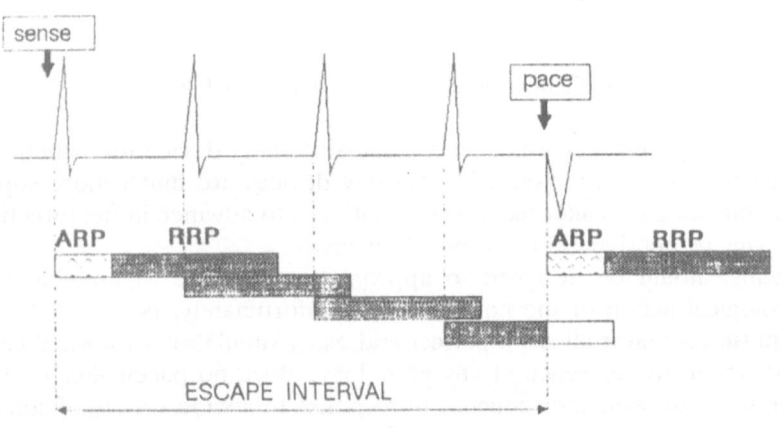

Figure 7. Underdrive pacing or dual demand mode. ARP = absolute refractory period; RRP = relative refractory period.

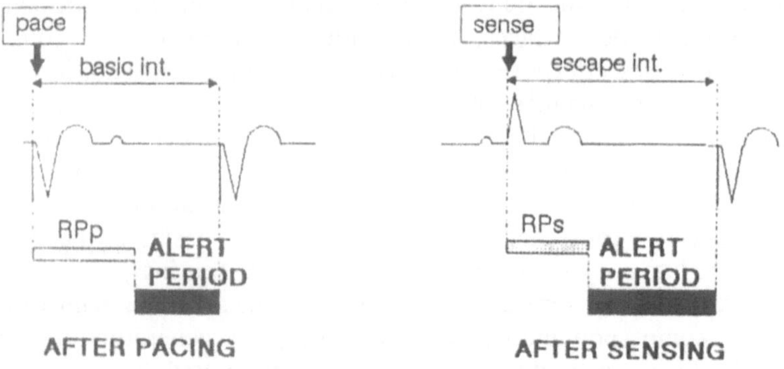

Figure 8. Alert interval of a VVI pacemaker after pacing and after sensing.

pacemaker to revert to the programmed basic rate. In some other systems, a short absolute refractory period (e.g. 125 ms) is followed by a relative refractory period of normal length; this concept can eventually be used for termination of ventricular re-entry tachycardias (*underdrive* pacing or *dual demand mode*) (Figure 7).

Alert interval:
The alert interval is the period of time during which the sensing amplifier is fully able to detect spontaneous heart activity, causing an adequate response of the output circuit. The alert interval starts upon completion of the refractory period and terminates at the end of the basic interval (or at the end of the escape interval if there is hysteresis) (Figure 8).

PHYSIOLOGICAL CARDIAC PACING

At first, the pacemaker was a simple life-sustaining device for patients with cardiac conduction problems. Present-day devices are much more sophisticated and still pacemaker technology continues to advance in the direction of improving responsiveness to the body's needs.

Pacing should be designed to approximate as close as possible to the physiological action of the normal heart. Unfortunately, no currently available pulse generator allows a perfect and exact simulation of normal cardiac physiology in the presence of any pathology. Since no pacemaker is able to mimic the heart rate and sequence in response to a wide variety of stimuli, it must be clear that physiological pacing seeks only to approximate more closely to normal physiology.

Physiological pacing is usually defined as artificial stimulation of the heart with a device which:

* *maintains the normal activation sequence of the heart (i.e. maintains A-V synchrony),*
* *restores the heart's ability to alter its rate physiologically.*

True physiological pacing has some distinct advantages over the simple ventricular pacing (VVI):

* as cardiac output depends strongly upon heart rate, P wave triggered atrial synchronous ventricular pacing may considerably augment the exercise tolerance of physically active patients whose sinus function is intact,
* due to a better ventricular filling (atrial boost effect), synchronization of atrial and ventricular contractions has been reported to produce an important increase in cardiac output compared with ventricular pacing at the same rate,
* at an equivalent cardiac output, physiological paced patients will almost certainly have lower end-diastolic pressures than those with ventricular pacemakers,
* A-V synchrony prevents cannon waves and reduces the potential for supraventricular and for ventricular tachyarrhythmias.

In an attempt to maintain or to restore the natural physiology of the heart, several principal types of pulse generators have been developed by various manufacturers:

* atrial demand pacemakers (AAI),
* atrial synchronous pacemakers (VAT & VDD),
* A-V sequential pacemakers (DVI),
* fully automatic pacemakers (DDD).

AAI pacemaker:

The AAI pulse generator is an on-demand device which is very similar to the VVI type: it paces and senses in the atrium only and its output circuit will be inhibited from discharging when spontaneous atrial activity occurs. This mode of pacing can be applied only in the *presence of normal, healthy A-V conduction.* The most often cited indication for atrial on-demand pacing is the 'sick sinus syndrome' or 'SSS' (i.e. disorders of the sinus node including sinus bradycardia, sinus arrest, and brady-tachy phenomena). However, in the presence of intact A-V conduction, AAI pacing may also be preferred for overdrive suppression of ventricular arrhythmias since it maintains cardiac output through A-V synchrony.

VAT pacemaker:

The VAT pacemaker, usually called P wave synchronous or P wave triggered, needs two leads. An atrial depolarization is sensed through the atrial lead and after an appropriate delay, corresponding to the normal P-R interval, the ventricle is paced via the ventricular lead. Such a system, sensing the natural P wave in the atrium and sequentially delivering a stimulus in the ventricle, is the equivalent of an antegrade-only atrioventricular bypass tract, in other words, it functions as an electronic AV-node and bundle of His. Obviously, the VAT system has applications in the presence of A-V block but does rely on *normal impulse formation in the atria* for optimum operation. A significant shortcoming of the VAT pacing mode is that spontaneous ventricular activity is either not sensed or falsely interpreted as an atrial depolarization. In that case, a ventricular stimulus may be delivered within the vulnerable period creating the potential danger of triggering an arrhythmia.

VDD pacemaker:
The VDD pulse generator is a P wave synchronous, ventricular inhibited pacemaker. Actually, it is an evolutionary version of the VAT type with an additional sensing amplifier for the ventricle. The VDD device is only pacing in the ventricle, but has dual chamber sensing and dual mode of response, i.e. an atrial sense triggers a ventricular stimulus and a ventricular sense inhibits the ventricular spike. As pacing in the vulnerable period is avoided by the ventricular sensing, there is no more risk for starting an arrhythmia, but the adequate functioning of the device still depends upon the normal impulse formation by the sinus node.

DVI pacemaker:
This type of pacemaker will pace in both chambers, but senses only in the ventricle and works only in the inhibited mode. In the absence of any ventricular activity, sequential stimuli will be delivered to both the atrium and the ventricle; this is the reason why this type of pulse generator is better known as the A-V sequential pacemaker. DVI pacing does not rely on normal A-V conduction (unlike the AAI types) nor does it depend upon normal impulse formation (as is necessary for VAT and VDD). As a consequence, the DVI pacemaker may have applications in the presence of A-V block and can also be applied to most conduction disorders. However, there is an important drawback: *atrial depolarizations have no effect on pulse generator function.* Therefore, the DVI pacemaker could potentially compete with atrial activity, which could conceivably produce atrial arrhythmias. Moreover, since atrial activity cannot be sensed and ventricular stimuli cannot be triggered, the ventricle can only be stimulated at the basic rate, even during exercise with a properly working sinus node.

DDD pacemaker:
The essence of DDD pacing is the maintenance of atrioventricular synchrony under most circumstances. The DDD pacemaker closely replicates the normal function of the sinus node and the A-V node of the heart and, therefore, it is often called *A-V universal* or also *fully automatic.*

The DDD pacemaker may be seen as a combination of the atrial synchronous device (VAT & VDD) and the A-V sequential device (DVI). When normal atrial depolarizations occur at a rate between the programmed upper and lower rate limits, ventricular pacing will be synchronized to this atrial activity, just as in the VDD pulse generator. However, when the spontaneous atrial activity decreases to a rate below the programmed lower rate limit, the atrium and the ventricle will be paced A-V sequentially, as in the DVI pulse generator (of course, there is only pacing in the ventricle when no normal A-V conduction occurs before the end of the programmed A-V interval). The DDD pacemaker is only one step away from being a physiological pacing system under all circumstances: it fails to adapt the heart rate to the metabolic demands if the sinus node doesn't work properly.

The DDD pacemaker is the *universal* device and most of the other pacing systems can be obtained by switching off one or more functions of this universal system. A better understanding of the principles of the DDD pacemaker may also give a clear insight into the working of the other types.

ESSENTIAL PARTS AND POSSIBILITIES OF A DDD PACEMAKER

The DDD pulse generator is a true *dual chamber* system and, thus, it requires two leads: one to pace the atrium and to sense spontaneous atrial activity and another to stimulate the ventricle and to detect intrinsic ventricular activity. If unipolar leads are used, the titanium can, which hermetically seals the electronic circuitry and the battery, acts as the indifferent electrode for both leads.

As illustrated in Figure 9, the DDD pacemaker is a double device: it is comprised of two separate output circuits and two distinct sensing units, sharing a common timing block and a common power source (for simplicity, the battery is not depicted in the figure).

The function of both atrial and ventricular *output circuits* is identical with the well-known circuit of the VVI pacemaker: it is still the output circuit with its output capacitor which is determining the shape and the amplitude of the pacing impulses.

The configurations and the purposes of the two *sensing circuits* are also similar to those of the VVI pacemaker. However, it should be noted that the sensitivity of the atrial channel has to be a little higher than that of the ventricular channel, because the intracardiac potentials measured in the atria are typically smaller than those obtained in the ventricles. There may also be a subtle difference in the bandpass characteristic of the filter in the atrial channel to accommodate the typically lower frequency components of the P waves. The refractory periods of the two sensing circuits are substantially different from each other and from the simple VVI device and, therefore, the refractory intervals are separately treated in the next section.

Although the clock oscillator in the *timer* is commonly shared by both the

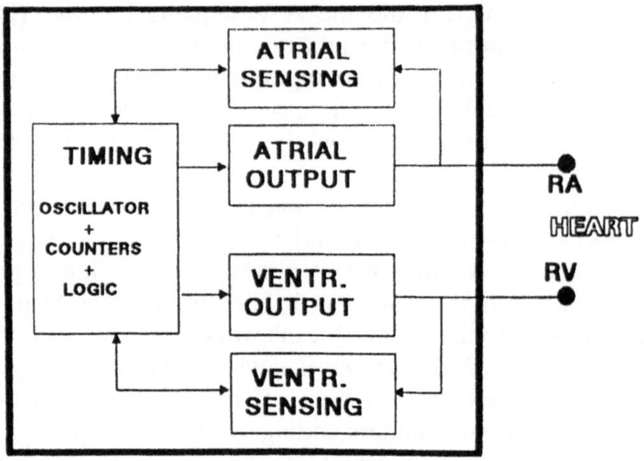

Figure 9. Block diagram of the DDD pacemaker. RA = right atrium, RV = right ventricle.

atrial and the ventricular channels, several distinct counters are needed for the determination of a multitude of different intervals (e.g. V-A escape interval, A-V delay, etc.). Moreover, since the atrial and the ventricular parts of the pulse generator have to work properly synchronized and without disturbing each other, some electronic circuitry is indispensable for coordination and control. These electronic circuits are known as *digital logic.*

According to the generic code, a DDD pacemaker must have the possibility to sense and to pace in both the atrium and the ventricle, and it will respond either in the triggered or in the inhibited mode. More exactly: while the atrial stimulus can only be blocked or inhibited by an atrial sensed event, the ventricular stimulus may be inhibited by a ventricular sensed event but can also be synchronized to (or triggered by) an atrial depolarization.

The operation of the DDD pacing mode with its four basic possibilities is illustrated in Figure 10.

A pace-V pace: A-V sequential pacing at lower rate

No atrial activity occurs within the programmed V-A escape interval and, consequently, an atrial output stimulus is delivered which is followed by an appropriate atrial response. Again, since no ventricular activity is sensed during the programmed A-V interval, a stimulus is delivered to the ventricle in order to produce a paced QRS complex.

A pace-V sense: atrial pacing with normal A-V conduction and hence ventricular inhibition

No electrical activity is sensed before completion of the V-A escape interval

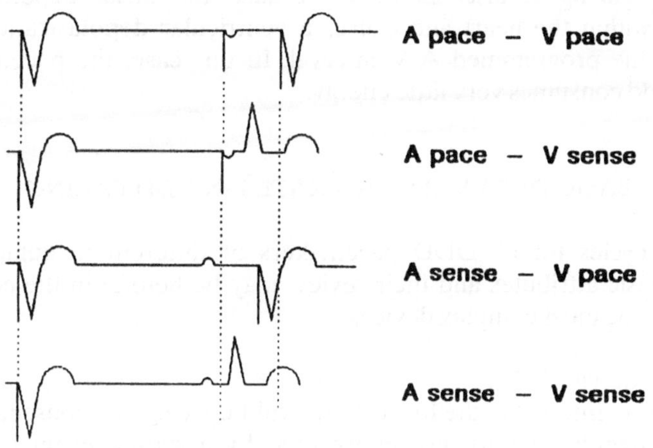

Figure 10. The four basic possibilities of DDD pacing. A = atrium; V = ventricle.

and, thus, a stimulus is launched into the atrium followed by an atrial depolarization. However, a spontaneously conducted ventricular contraction occurs before the end of the programmed A-V interval and inhibits the ventricular discharge. In this situation, the patient's heart rate will be slightly faster than the programmed lower rate (at least for one beat; as a matter of course continuously pacing in the atrium followed by spontaneous conduction is always at basic rate).

A sense-V pace: atrial synchronous ventricular pacing
A spontaneous P wave occurs before the end of the programmed V-A escape interval and, therefore, the atrial output is blocked. The sensed atrial contraction starts the A-V interval counter and in the absence of a naturally conducted ventricular depolarization, a stimulus is delivered to the ventricle at the end of this programmed A-V interval. The pacing impulse is further followed by a paced QRS complex and instantly a new V-A escape interval is started. Obviously, the ventricular pacing rate responds by increasing and decreasing with the atrial rate, as long as the patient's fluctuating atrial rate remains between the programmed lower and upper rate limits. Within these limits, the DDD pacemaker *automatically* adapts to the rhythmogenic needs of the heart.

A sense-V sense: stand by with spontaneous P wave and normal A-V conduction
The atrial pacing output is inhibited, since a normal P wave occurs before completion of the programmed V-A escape interval (e.g. when the patient's own spontaneous rate is faster than the programmed lower rate limit).

Ventricular pacing is also inhibited because the atrial depolarization is conducted within the heart and causes a ventricular depolarization prior to the end of the programmed A-V interval. In this case, the pacemaker is in 'stand by' and consumes very little energy.

BASIC INTERVALS AND RATES IN DDD PACING

All timing cycles for all DDD pacemakers of different manufacturers are similar in basic attributes and their review may be helpful in the comprehension of even the most complex devices.

Lower rate interval (LRI):
The lower rate interval is the longest interval between two consecutive paced ventricular depolarizations, not interrupted by a sensed event (Figure 11). For the DDD type, the lower rate interval is the equivalent of the *basic interval* in VVI pacemakers.

In the absence of hysteresis, the lower rate interval is also the longest period of time between a sensed spontaneous ventricular event and the following paced event. In this situation, which is true for many DDD trademarks, the lower rate interval (LRI) equals the *ventricular escape interval.*

Lower rate limit (LRL):
The lower rate limit is the programmed minimum rate at which the pacemaker will pace in both chambers, provided that the intrinsic atrial rate is slower than that rate and in the absence of adequate A-V conduction.

The main objective of pacemakers has always been the prevention of dizziness and syncope. Therefore, the lower rate limit is still essential in the management of bradyarrhythmias.

The lower rate limit is usually expressed in beats per minute (bpm), while the lower rate interval is measured in milliseconds (ms). Since one minute equals 60 000 milliseconds, the relation between lower rate interval (LRI) and lower rate limit (LRL) is given by

$$\text{LRI (ms)} = \frac{60\,000}{\text{LRL (bpm)}}$$

and also

$$\text{LRL (bpm)} = \frac{60\,000}{\text{LRI (ms)}}$$

Escape interval:
In a VVI pulse generator, the escape will occur by a ventricular stimulus, but in a DDD pacemaker, escape will occur by an atrial stimulus. Hence, for DDD devices, the escape interval means the longest period of time between a

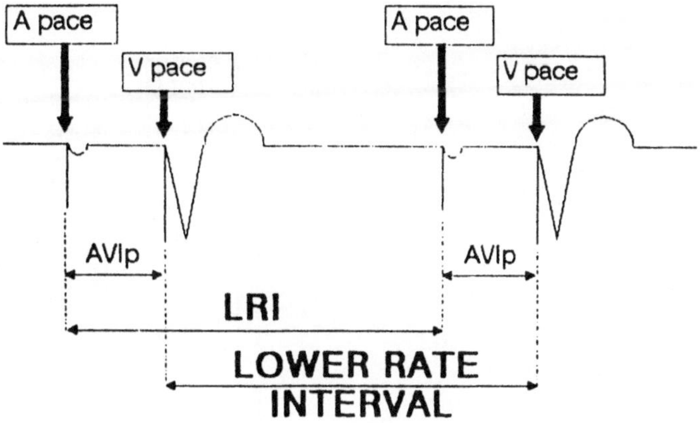

Figure 11. Definition of lower rate interval (LRI). AVIp = programmed atrioventricular (A-V) interval.

sensed spontaneous atrial depolarization and the subsequent paced atrial event.

In the absence of an adequate A-V conduction, the escape interval is also the longest interval between the ventricular stimulus following after a sensed spontaneous atrial contraction and the next ventricular stimulus following after a paced atrial event (Figure 12).

Escape rate:
The escape rate is the programmed lowest stimulation frequency of the pacemaker below which the ventricular rate impossibly can fall.

Since the escape rate expressed in beats per minute (bpm) and the escape interval measured in milliseconds (ms) are the inverse of each other, it follows in the same way as for the lower rate limit:

$$\text{Escape Rate (bpm)} = \frac{60\,000}{\text{Escape Interval (ms)}}$$

Note that no difference exists between the lower rate limit (LRL) and the escape rate when the hysteresis function is switched off, or isn't available for a particular make of DDD pacemaker.

Upper rate interval (URI):
The upper rate interval is the shortest possible interval between two consecutive ventricular stimuli or between a sensed intrinsic ventricular depolarization and the following paced ventricular event (Figure 13).

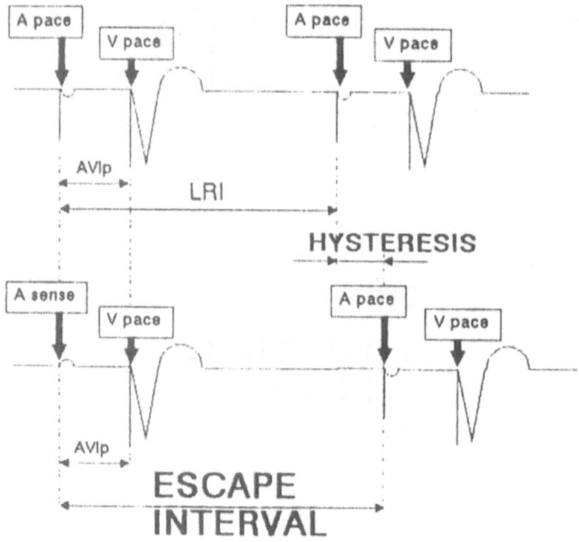

Figure 12. Definition of escape interval. LRI = lower rate interval; AVIp = programmed A-V interval.

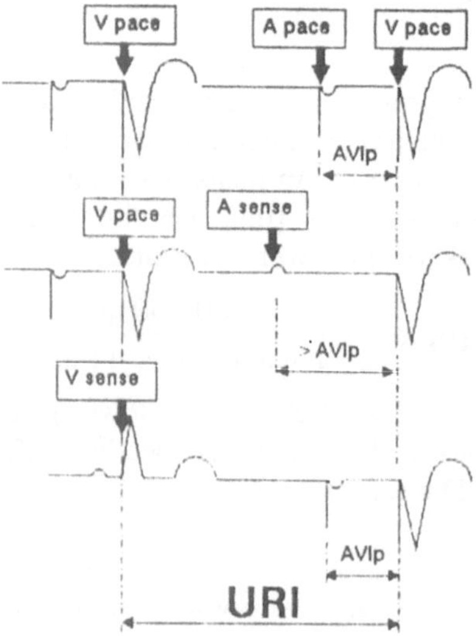

Figure 13. Definition of upper rate interval (URI). AVIp = programmed A-V interval.

Upper rate limit (URL):
This is the programmed maximum ventricular paced rate, i.e. the highest frequency of ventricular depolarizations induced by the pacemaker.

The relation between upper rate limit (URL) in beats per minute and upper rate interval (URI) in milliseconds is easily found:

$$\text{URL (bpm)} = \frac{60\,000}{\text{URI (ms)}}$$

Some precautions are required since a DDD pulse generator senses in the atrium and may deliver impulses to the ventricle at the same rate. Therefore, the upper rate limit is mandatory to avoid a dangerous ventricular response to excessively rapid atrial events such as atrial fibrillation or flutter. The upper rate limit also protects against potentially dangerous influences from the surroundings (inside or outside the body) such as electromagnetic interference (EMI) or even myopotentials.

Although all pacemaker manufacturers enforce the most rigorous standards in the design and manufacture of their pacemakers, there is nonetheless a statistically finite possibility of malfunction. Therefore, pacemakers are protected against the remote feasibility of a pulse rate *runaway* by an independent electronic circuit which limits the pulse rate. So pay attention to the difference: the upper rate limit is programmable, while the absolute maximum of the runaway protection is always fixed by the manufacturer.

A-V interval (AVI):
The atrioventricular interval (AVI) is the period of time elapsing between an atrial paced or sensed depolarization and the succeeding ventricular stimulus. It is the electronic equivalent of the P-R interval of the heart.

Some pacemakers have a shorter atrioventricular interval after a sensed atrial event (AVIse) than after an atrial pacing stimulus (AVIpa) (Figure 14). The reason for this difference is to provide an equal filling time for the ventricle in both cases. In fact there is always a certain delay between the atrial stimulating impulse and the spread-out of the depolarization over the atrial walls; moreover it will also take some time for the depolarization originating from the sinus node to propagate to the site of the atrial sensing electrode. To compensate for both small delays the A-V interval after an atrial stimulus has to be somewhat longer than the one after sensing.

The A-V interval is also known as the *A-V Delay* and is always programmable:

* for each particular patient, the cardiac output may be optimized by adjusting the A-V interval,
* a short A-V interval might depolarize the normal as well as the accessory conduction pathways in the retrograde direction and, thus, prevent the establishment of a re-entry tachycardia.
* in some cases (e.g. first and second degree AV block and intermittent

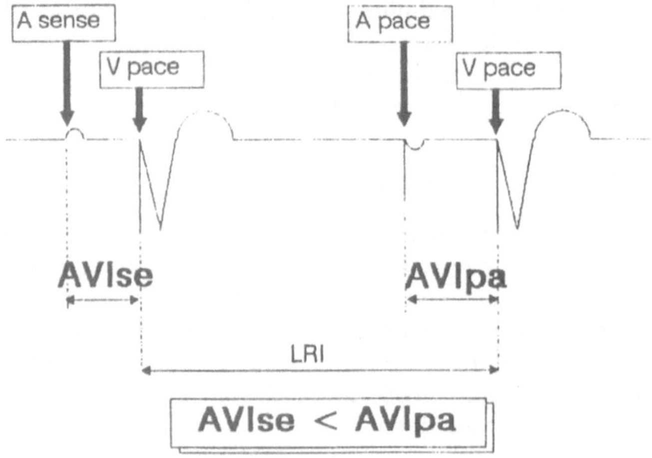

Figure 14. Definition of atrioventricular interval (AVI). AVIse = A-V interval after atrial sensing; AVIpa = A-V interval after atrial pacing; LRI = lower rate interval.

complete AV block), a long atrioventricular interval allows enough time for the heart to become depolarized through the normal conduction system; a normal spread-out of the ventricular depolarization can be of diagnostic value and, at the same time, will save energy and extend the battery life,

* by programming a rather long A-V interval, the pacemaker can be used as an AAI pulse generator with the ventricular channel in stand-by mode (i.e. ventricular back-up rate).

V-A interval (VAI):
The ventriculo-atrial interval (VAI) is frequently measured on the ECG strip, since it is very helpful in the analysis of the normal DDD pacemaker operation. It always starts with a ventricular event, paced or sensed, and is terminated by an atrial stimulus.

In most cases, *when no hysteresis is present*, the programmed ventriculo-atrial interval (VAIp) is the longest period of time between a sensed or paced ventricular depolarization and the succeeding atrial stimulus. Obviously, the V-A interval may then be calculated by subtracting the A-V interval from the lower rate interval (Figure 15):

$$\text{VAIp} = \text{LRI} - \text{AVIp}$$

However in DDD *devices with hysteresis*, the V-A interval can be longer than calculated in the above-mentioned formula if the ventricular event is pre-

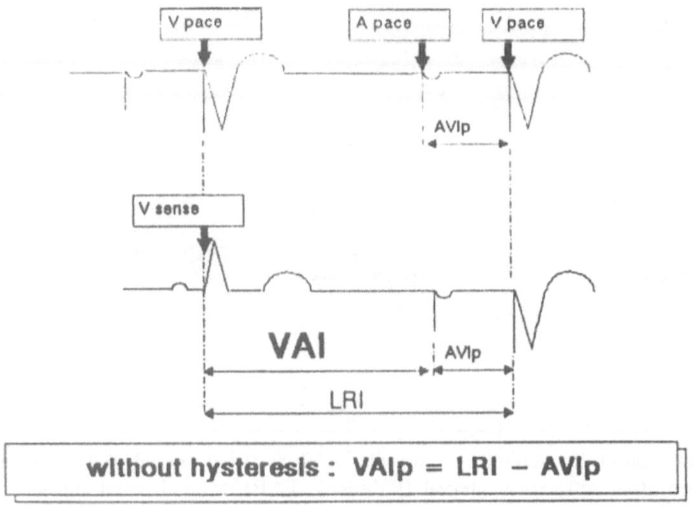

Figure 15. Definition of V-A interval (VAI). VAIp = programmed V-A interval; AVIp = programmed A-V interval; LRI = lower rate interval.

ceded by a sensed P wave. In that case, the lower rate interval should be replaced by the escape interval.

INSENSITIVE INTERVALS AND CROSSTALK PROBLEMS

As for the VVI pulse generator, each channel of a DDD pacemaker needs its own refractory period during which the sensing circuit is switched off. However, DDD pacemakers being real 'dual chamber' devices which are able to sense in both the atrium and the ventricle, may also inappropriately sense electrical activity that is not intended to be sensed. For example, the ventricular sensing amplifier is designed to sense the spontaneous ventricular electrical activity, but it may also sense atrial output impulses when no extra precautions are taken.

Ventricular refractory interval (VRI):
The ventricular refractory interval is the period of time following either a sensed or a paced ventricular event during which the pacemaker will not sense any electrical activity.

In most of the present-day DDD pacemakers, the ventricular refractory interval (VRI) consists of two consecutive parts: an absolute refractory interval followed by a relative refractory interval (Figure 16). The *absolute*

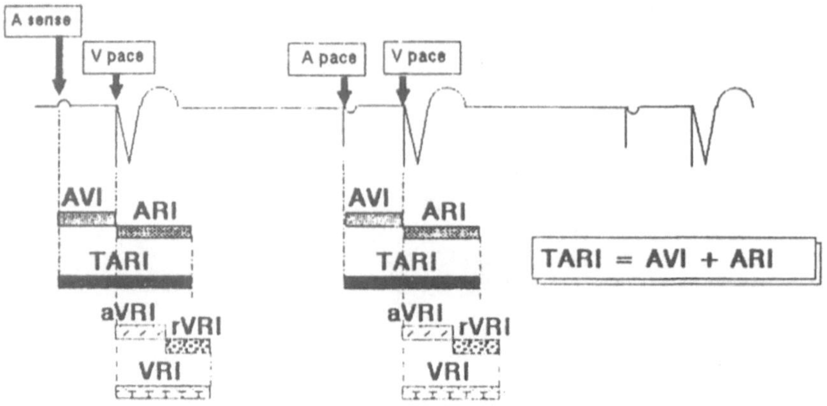

Figure 16. Definitions of refractory intervals. AVI = programmed A-V interval; ARI = postventricular atrial refractory interval (PVARP); TARI = total atrial refractory interval; VRI = ventricular refractory interval; aVRI = absolute refractory part of VRI; rVRI = relative refractory part of VRI.

refractory interval is the first portion of VRI during which the detection of all signals is blocked. The *relative refractory interval,* also called *noise sampling interval,* is the second part of VRI during which detected signals are evaluated for repetition rate. A ventricular sense during this noise sampling period retriggers the ventricular refractory interval (the whole VRI or only the relative refractory part, depending upon the pacemaker make). Continuous retriggering of the VRI by a ventricular tachycardia, myopotentials or electromagnetic interference (EMI) causes the pulse generator to function asynchronously at a lower rate. During a rapid tachycardia and with the pacemaker no longer sensing, a pacemaker spike will eventually fall into the 'terminating window' and terminate the tachycardia. This method of underdrive pacing is known as the *dual demand mode* (see also Figure 7).

For underdrive pacing in the dual demand mode, a QRS complex has to be sensed at least once during each newly started ventricular refractory interval (VRI). Therefore, for terminating a tachycardia with a rate of 200 bpm, a minimum ventricular refractory period (VRI) of 300 ms is needed.

Atrial refractory interval (ARI):
The atrial refractory interval (ARI) is the interval after a ventricular paced or sensed depolarization during which the atrial channel is unable to detect any incoming signal. Since the atrial refractory interval (ARI) always starts with a ventricular event, it is often called the *postventricular atrial refractory period* and abbreviated as *PVARP.* In most present-day DDD pacemakers, the ARI is programmable but, for some trademarks, it can also be fixed.

The main objectives of the atrial refractory interval (ARI) are twofold:
* inappropriate atrial sensing of ventricular events (ventricular stimuli, QRS complexes, and abnormal T-waves) has to be avoided,
* atrial sensing of retrogradely conducted P waves or aberrant atrial events must be eliminated.

It should be noted that the length of the atrial refractory interval also limits the brevity of the allowable interval between two sensed atrial events; so another supplementary objective of ARI might be the limitation of the upper rate of DDD pacemaker response.

Total atrial refractory interval (TARI):
The total atrial refractory interval is the period of time after an atrial sensed or paced event during which all electrical signals at the input of the atrial channel will be ignored.

It is obviously undesirable for a second atrial event to be sensed once the A-V interval has begun. Therefore, during the atrioventricular interval (AVI), the atrial channel is always refractory. At the end of the A-V interval, a ventricular stimulus is launched and a postventricular atrial refractory interval (ARI) is started.

So, the total atrial refractory interval (TARI) always consists of two consecutive portions: it is made up of the sum of the atrioventricular interval (AVI) and the postventricular atrial refractory interval (ARI) (Figure 16). In formula:

$$\text{TARI (ms)} = \text{AVI (ms)} + \text{ARI (ms)}$$

It follows also that TARI, as a whole, is always programmable. In some devices, the ARI is fixed and only the A-V interval can be programmed. In other pacemaker makes, the TARI and the AVI may be chosen separately.

In some DDD pacemakers, the ventricular refractory interval (VRI) and the postventricular atrial refractory interval (ARI) are coupled, i.e. the ARI is a little bit longer (e.g. 30 ms) than the absolute refractory part of VRI. During the relative refractory part of VRI, the ventricular sensing is already restored in some way. This helps to prevent a ventricular event being sensed by the atrial channel without being seen by the own ventricular channel and, thus, avoid the incorrect start of an A-V interval.

Crosstalk:
Crosstalk is a well known notion of the electronics such as telephone, radio, and many other data communication systems. It refers to the interference due to mutual coupling between two adjacent circuits, producing an unwanted signal in one circuit when a signal is present in the other. *Applied to the world of DDD pacemakers, crosstalk is the inappropriate detection of an event in one chamber by the sensing amplifier of the other chamber, resulting in a derangement of timing and pacing.*

Depending upon the origin of the signals, crosstalk is classified as *far-field sensing* or as *self-inhibition.* Both aspects of crosstalk are treated separately in the following sections.

Far-field sensing:
Crosstalk is classified as *far-field sensing* when the unwanted signals are due to activity of the heart muscle.

Electrical fields caused by the action potentials of the heart cells, extend through the body tissue and, thus, voltages may be measured between an electrode inside the heart and the can of the pacemaker which is used as an indifferent electrode. The field strength decreases rapidly with the distance from the depolarizing cells and so does the amplitude of the detected voltages. However, when a large number of cells act in synchrony and fire nearly in unison, their individual signals will be added to a larger voltage. The atrial mass is normally too small to produce a disturbing voltage at the site of the ventricular electrode. On the contrary, when the large number of cells of the thick ventricular walls depolarize simultaneously, a considerable voltage is generated which eventually may be observed at the site of the atrial electrode.

From these considerations it can be concluded that for dual chamber pacemakers, the definition of *far-field sensing* may be restricted as follows: *Far-field sensing is the erroneous atrial sensing of ventricular signals, either QRS complexes or PVC's, resulting in the inappropriate triggering of a ventricular output impulse at the end of a programmed A-V interval.*

The far-field sensing was a real problem with the old VAT pulse generators. Sensing of a QRS complex followed by a new ventricular pacing stimulus after an A-V delay equal to the programmed AVI, was able to start a kind of pacemaker-induced tachycardia at a rate of nearly 60 000 divided by AVI.

Far-field sensing only occurs on rare occasions with present-day DDD pacemakers. Remaining problems can always been solved by programming a postventricular atrial refractory interval (ARI) which is slightly longer than the absolute part of the ventricular refractory interval (aVRI).

Self-inhibition:
Crosstalk is classified as 'self-inhibition' when the unwanted signals are not originating in the myocardial cells but are due to the 'hardware' of the pacemaker system.

As soon as a ventricular pacing stimulus is started, the atrial channel is made refractory (cf. ARI). It is obvious that neither the ventricular spike nor the 'afterpotentials' following immediately after the impulse, can influence the atrial sensing circuit. The problems of sensing unwanted hardware signals are thus restricted to the ventricular channel. Hence the definition: *Self-inhibition is the inappropriate ventricular sensing of the atrial stimulus artifact or related polarization potentials, resulting in intermittent or complete inhibition of the ventricular output.*

The prevention of self-inhibition is perhaps the most important challenge in designing and building modern dual chamber pacemakers such as DDD pulse generators. Several solutions have been developed by different manufacturers and may be used alone or in combination:

* committed pacing,
* ventricular blanking,
* fast recharge impulses,
* nonphysiologic A-V delay and ventricular safety pacing.

These different approaches are treated in the next sections.

Committed pacing:
In this mode of pacing, a ventricular stimulus is imperatively linked (i.e. committed) to the atrial stimulus. This means that an atrial stimulus will *always* be followed by a ventricular stimulus: even a spontaneously conducted QRS or a PVC during the atrioventricular interval (AVI) following the atrial spike will no longer inhibit the ventricular output circuit (Figure 17).

Committed pacing is the simplest but least elegant method to avoid self-inhibition. It is obtained very easily by starting the ventricular refractory period at the atrial stimulus, i.e. after delivery of an atrial spike the ventricular channel is made refractory during the A-V interval (AVI) and for the following ventricular refractory interval (VRI).

Long ventricular pauses due to self-inhibition are circumvented by committed pacing, but as an atrial stimulus always forces a ventricular spike after the A-V interval, a committed device is readily competitive with spontaneous ventricular activity.

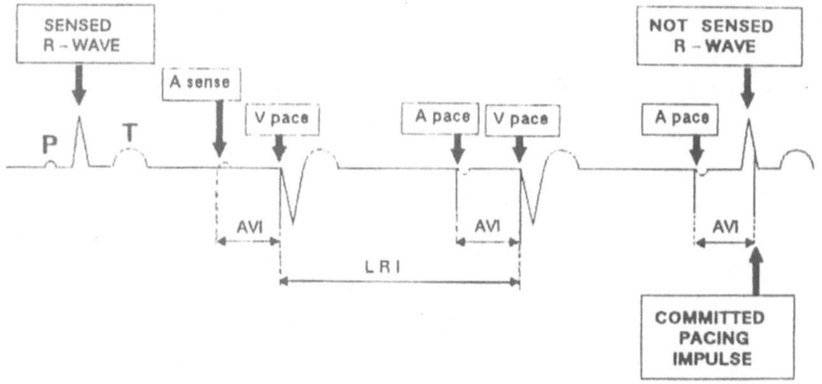

Figure 17. Principle of committed pacing. AVI = programmed A-V interval.

Ventricular blanking interval:
The ventricular blanking period stands for a short and temporarily disabling of the ventricular sensing circuit, starting at the atrial stimulus. It avoids the sensing of the atrial stimulus in the ventricular channel (Figure 18).

Blanking intervals and *refractory intervals* are both insensitive portions of the pacemaker cycle during which the sensing circuit is incapable of recognizing any electrical activity. However, the *blanking interval* is short and is designed to avoid detection of hardware activity (i.e. stimuli and related after potentials), while the *refractory interval* is much longer and avoids the sensing of physiological activity (i.e. QRS complexes and T waves).

If the blanking period is too long, there is a possibility that naturally occurring R waves coincide with the blanking period and will fail detection, resulting in a competitive ventricular output impulse. If the blanking period is too short, there is still an opportunity for self-inhibition. Older DDD pacemakers were equipped with fixed and rather long blanking intervals (e.g. 60 or even 75 ms); most present-day pulse generators have much shorter blanking intervals (e.g. 12 to 20 ms) and, for some devices, the blanking is programmable or coupled with the impulse duration (see also afterpotentials in the next section).

Afterpotentials and fast recharge pulses:
During the delivery of a cathodal pacemaker impulse, positively charged ions (e.g. Na^+) accumulate at the electrode-heart interface. The negatively charged metal electrode and the positively charged ions in the electrolyte are both surrounded by water molecules, forming a diffuse double layer which is a barrier for charge transfer (Stern, 1924). The effect of this charged bilayer is

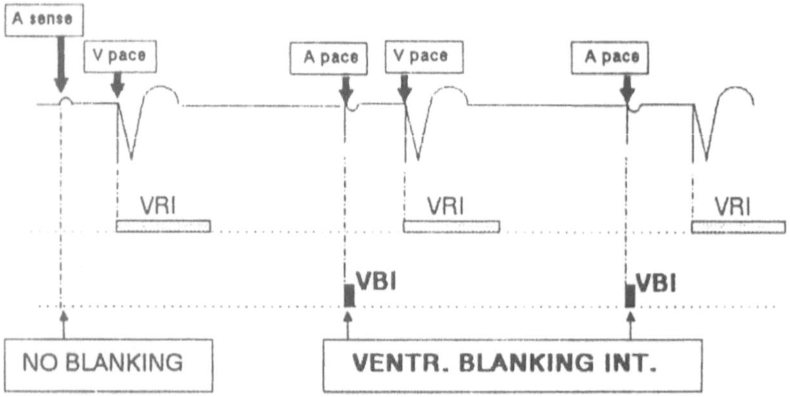

Figure 18. Definition of ventricular blanking interval. VRI = ventricular refractory interval; VBI = ventricular blanking interval.

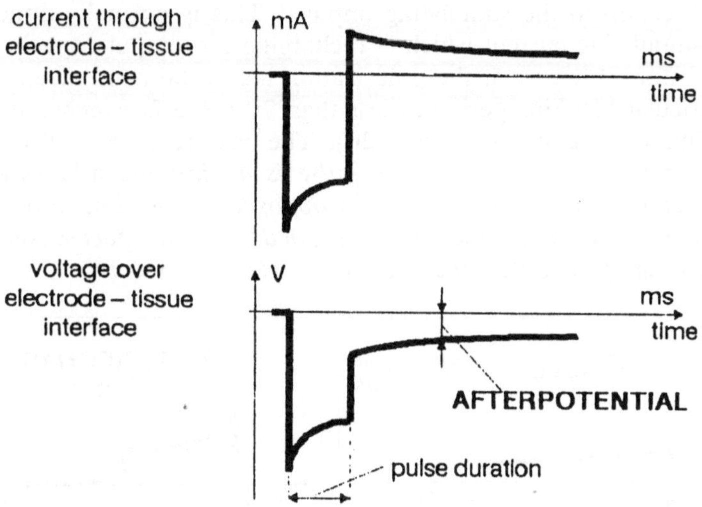

Figure 19. Afterpotential due to polarization.

that of a capacitor and is therefore called the Helmholtz capacitance C_H. At high current densities, oxido-reduction reactions will create an ohmic pathway parallel to C_H and this is normally referred as the Faradaic resistance R_F.

When a current is passed through the electrode, the capacitor C_H will accumulate a charge and a voltage is built up. This whole phenomenon is known as *polarization* and the voltage, which is opposing the pacing impulse, is called the *polarization voltage.* As it takes some time for the accumulated ions to be dispersed again, the polarization voltage is also present at the electrode surface for quite some time after the stimulating impulse. Hence, the definition of afterpotentials (Figure 19): *The afterpotential is the slowly decreasing polarization voltage at the electrode-tissue interface following the delivery of a pacemaker stimulus.*

Obviously, polarization voltages will arise at both the electrodes, i.e. at the ring and the tip with bipolar systems and at the tip and the can of the pacemaker with unipolar syostems. Since in unipolar systems, the pacemaker can is a common electrode for both the atrial and the ventricular channel, the polarization voltage at the can caused by one channel may be sensed by the other channel.

The polarization voltage and, consequently, the afterpotentials augment with increasing pulse amplitude and pulse duration. Moreover, with increasing pacing rate there is less time between two impulses for the ions to dissipate and the afterpotentials are relatively larger. In this circumstances, the ventricular blanking interval may be increased to avoid self-inhibition.

Between two pacing spikes, the charge built up by polarization in the interface capacitance C_H may be reduced to zero by an electric current of opposite direction to the stimulating impulse. This is normally done by the small nonstimulating current which is recharging the output capacitor of the pulse generator. However, if the polarization has to be neutralized during a short ventricular blanking period of less than 20 ms, a fast recharge pulse of high amplitude is required (Figure 20). The fast recharge pulse may be defined as follows: *The fast recharge pulse is an impulse following shortly after the pacing spike and with a polarity opposite to the stimulating impulse. It is intended to recharge the output capacitor of the pacemaker and to eliminate the afterpotentials at the electrode in a few milliseconds.*

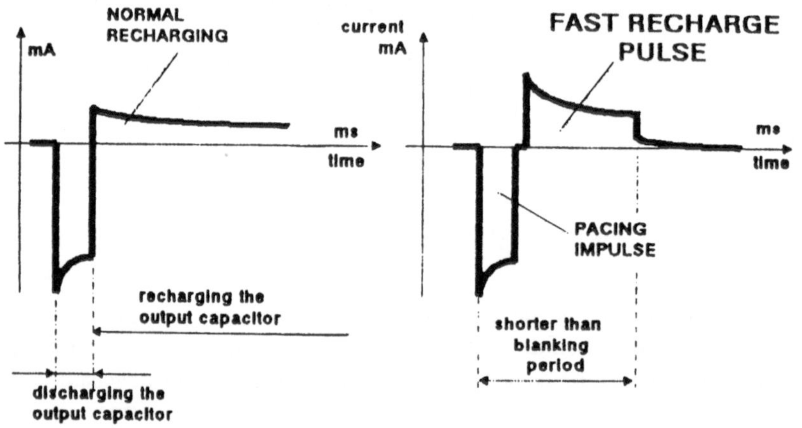

Figure 20. Difference between normal and fast recharging of the output capacitor.

Although the amplitude of some fast recharge pulses may be rather high, these impulses will never stimulate the myocardium, since they come immediately after a pacing impulse. Nevertheless, since fast recharge pulses neutralize the afterpotentials in the very short time of a blanking period, they enable the ventricular sensing circuit to detect eventual QRS complexes or PVCs following very close after the atrial stimulus.

Polarization potentials are rather small and slowly varying voltages which are normally blocked by the filter and the level detector of the sensing circuit. However, eventual interfering signals (EMI, myopotentials, etc.) may be summed up with the polarzation potentials and thus increase the likelihood of false inhibition.

Nonphysiologic A-V delay and ventricular safety pacing:
The *nonphysiologic A-V delay* is a special interval started by an atrial pacing impulse and during which the ventricular sensing circuit, although able to

detect electrical activity, will not react with an inhibition of the ventricular output circuit but, on the contrary, will trigger that circuit to deliver a ventricular pacing impulse at the end of this nonphysiologic A-V delay.

Not all DDD pulse generators are equipped with a nonphysiologic A-V delay, but if this special interval is available, it is always shorter than or equal to the normal programmed atrioventricular interval (AVIp). Usually, the nonphysiologic A-V delay has a duration of about 100 to 110 ms and it may be even programmable for some present-day DDD devices.

Intrinsic P-R intervals are usually longer than 100 ms, thus electrical activity sensed within the first 100 ms following an atrial pace, is in all probability caused by interference (e.g. EMI) or by electrode polarization signals, i.e. *non-physiologic.* Therefore, this first part of the A-V interval is called the *nonphysiologic A-V delay.* The inhibition of the ventricular channel by sensing of interference may have disastrous consequences for pacemaker-dependent patients. For those patients, a pacing impulse at the end of the nonphysiologic A-V delay is a *safety* precaution against possible interference and, therefore, this particular interval is also known as the *safety window.*

Ventricular safety pacing is to generate a ventricular output impulse at the end of a nonphysiologic A-V delay following an atrial stimulus, if during that interval some electrical activity was sensed by the ventricular sensing circuit (Figure 21). As already explained, ventricular safety pacing protects against inappropriate inhibition of the ventricular channel by EMI, myopotentials, etc.

Ventricular safety pacing is a restricted form of committed pacing. With

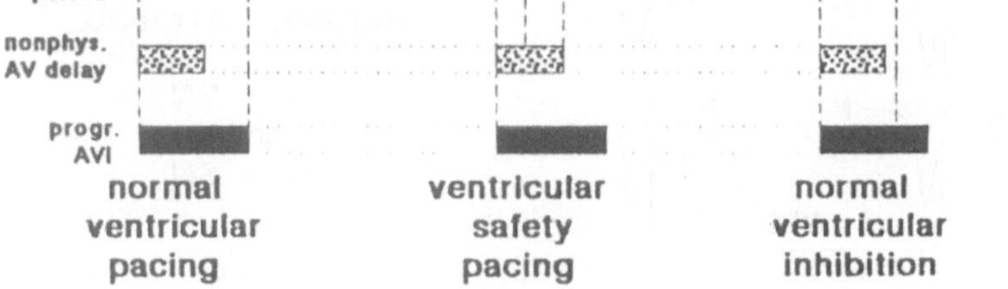

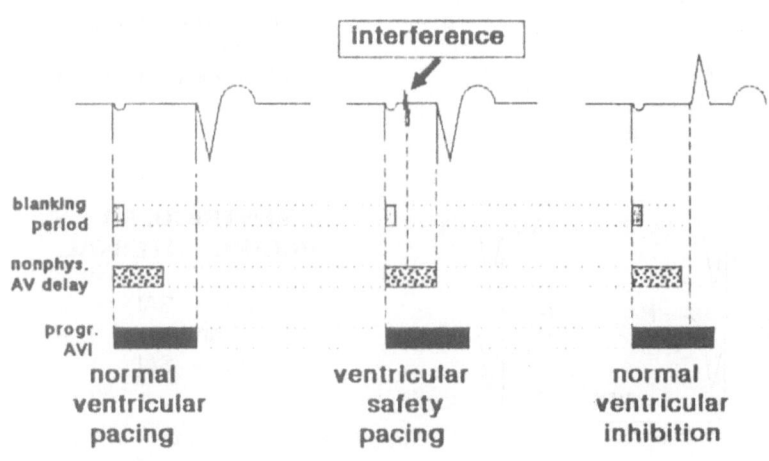

Figure 21. Definition of nonphysiologic A-V delay and principles of ventricular safety pacing.

the real committed pacing the ventricular sensing circuit is refractory during the whole programmed A-V interval. On the contrary, with the safety pacing the capability to inhibit the ventricular channel is restored after the first part of the programmed A-V interval.

A QRS complex or a PVC occurring during the 'safety window' will lead to a ventricular stimulus at the end of the nonphysiologic A-V delay. Since this stimulus will fall in the heart's refractory period, it will consequently have no effect.

VENTRICULAR TRACKING

Spontaneous atrial interval (SAI):

The spontaneous atrial interval (SAI) is the elapsed time between two consecutive spontaneous atrial depolarizations expressed in milliseconds (ms) (Figure 22).

As a matter of course, the spontaneous atrial interval is not at all determined by the pacemaker but depends upon the metabolic demands of the patient (at least for those active patients with AV block but intact sinus node function, being the ideal patients to receive a DDD pulse generator).

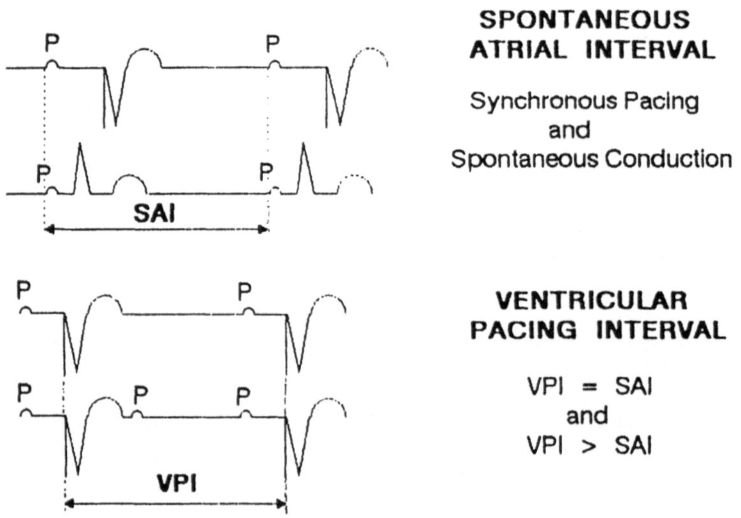

Figure 22. Definitions of spontaneous atrial interval (SAI) and ventricular pacing interval (VPI).

Spontaneous atrial rate (SAR):
The spontaneous or intrinsic atrial rate is the frequency of the atrial depolarizations without intervention of the pacemaker, i.e. the number of spontaneous P waves per minute. The relation between the spontaneous atrial interval (SAI) measured in ms and the intrinsic atrial rate (SAR) expressed in bpm is given by:

$$\text{SAR (bpm)} = \frac{60\,000}{\text{SAI (ms)}}$$

or

$$\text{SAI (ms)} = \frac{60\,000}{\text{SAR (bpm)}}$$

Ventricular pacing interval (VPI):
The ventricular pacing interval is the period of time, computed in ms, between two consecutive paced ventricular events without an intervening spontaneous ventricular depolarization (Figure 22). Note that the ventricular pacing interval is not necessary equal to the spontaneous atrial interval (see, for instance, upper rate behavior in the next section).

Ventricular pacing rate (VPR):
The ventricular pacing rate is the frequency of the ventricular depolarizations evoked by the pacemaker provided that there is no spontaneous ventricular activity, i.e. the number of paced ventricular events expressed in beats per minute when there is no underlying ventricular heart rhythm.

The connection between ventricular pacing rate (VPR) and ventricular pacing interval (VPI) follows from the well known expression:

$$\text{VPR (bpm)} = \frac{60\,000}{\text{VPI (ms)}}$$

Tracking:
The ventricle '*tracks*' the atrium, when each spontaneous P wave is followed by a ventricular paced event, i.e. when the ventricular pacing responds synchronously to the sensed atrial activity in a 1 : 1 way (Figure 23).

From this definition, it manifestly follows that during tracking the ventricular pacing interval equals the spontaneous atrial interval (VPI = SAI).

Since a DDD pacemaker is unable to stimulate the ventricle at rates beyond the programmed values of the lower rate limit (LRL) and the upper rate limit (URL), the tracking is restricted within these limits. Between these two end points the ventricular pacing rate (VPR) is directly proportional to the spontaneous atrial rate (SAR) and the graphic representation of VPR as a function of SAR is a straight line (see diagram of Figure 23). However, as soon as the intrinsic rate (SAR) falls below the lower rate limit (LRL),

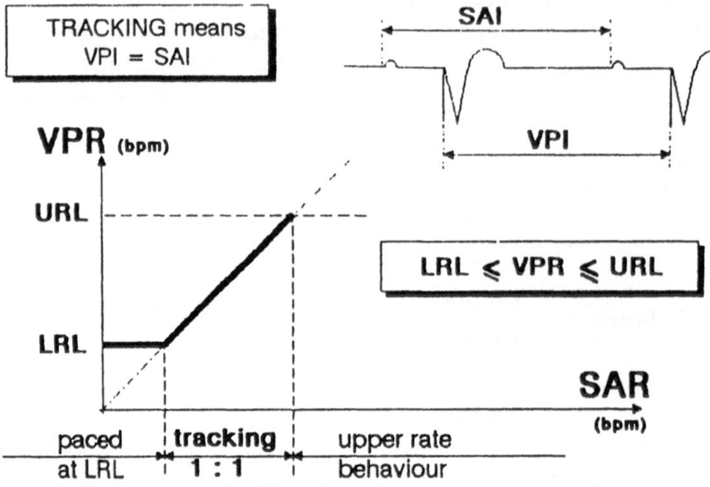

Figure 23. Definition of tracking. VPI = ventricular pacing interval (ms); VPR = ventricular pacing rate (bpm); SAI = spontaneous atrial interval (ms); SAR = spontaneous atrial rate (bpm); URL = upper rate limit (bpm); LRL = lower rate limit (bpm).

tracking becomes impossible and the pulse generator stimulates at the programmed lower rate limit. If, on the opposite end, the spontaneous atrial rate (SAR) gets higher than the upper rate limit (URL), the ventricle can also no longer 'track' the atrium. With those high atrial rates the behavior of the pacemaker will depend upon the different settings of URI and TARI and an answer to these problems is given in the next section.

Ventricular tracking limit (VTL):
The ventricular tracking rate is the maximum rate, expressed in beats per minute, at which the ventricle can be paced when it is responding to the atrial activity in a 1:1 way.

Since it is impossible for the ventricle to 'follow' or 'track' the atrium above the programmed upper rate limit (URL), the ventricular tracking limit (VTL) is just another name for the same concept, i.e. URL = VTL.

UPPER RATE BEHAVIOR OF DDD PACEMAKERS

Blocking onset rate (BOR):
The blocking onset rate is an important concept in understanding the upper rate behavior of DDD pacemakers. It is the maximum atrial rate which can

be sensed by the atrial sensing circuit without omitting or 'blocking' some P waves.

The ability of the atrial sensing circuit to detect continuously the spontaneous atrial activity is limited by the total atrial refractory period, i.e. the shortest detectable spontaneous atrial interval (SAI) has to be longer than the total atrial refractory interval (TARI) (Figure 24).

Since the blocking onset rate is expressed in beats per minute, while the total atrial refractory interval is measured in milliseconds, their relation is easily found:

$$\text{BOR (bpm)} = \frac{60\,000}{\text{TARI (ms)}}$$

Upper rate behavior:
Whenever the spontaneous atrial rate (SAR) exceeds the upper rate limit (URL) of a DDD pacemaker, the behavior of the device will be dictated by the programmed settings. Two different modes of response may be distinguished depending upon the relationship between the total atrial refractory interval (TARI) and the upper rate interval (URI):

* *Block response*: when TARI = URI or BOR = URL,
* *Pseudo-Wenckebach response*: when TARI < URI or BOR > URL.

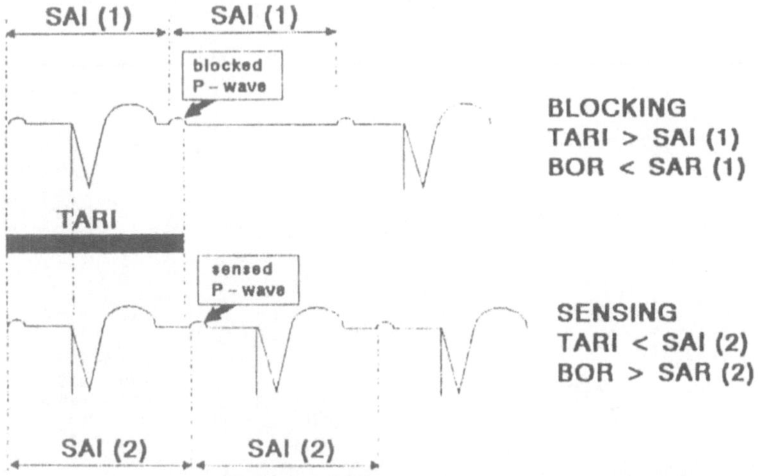

Figure 24. Definition of blocking onset rate (BOR). SAI = spontaneous atrial interval (ms); SAR = spontaneous atrial rate (bpm); TARI = total atrial refractory interval (ms) = AVI + ARI (sum of programmed atrioventricular interval and postventricular atrial refractory interval).

Both possibilities will be illustrated by means of an example in the two following sections:

Block response:
When the total atrial refractory interval (TARI) equals the upper rate interval (URI), block response will occur as soon as the spontaneous atrial interval (SAI) becomes shorter than the programmed upper rate interval (URI), i.e. when the spontaneous atrial rate (SAR) becomes higher than the upper rate limit (URL).

This situation is illustrated in Figure 25. The first P wave is detected and starts an A-V interval (AVI) at the end of which a ventricular spike is generated starting on its turn a postventricular atrial refractory interval (ARI). The second P wave falls within this atrial refractory interval, is therefore not seen by the atrial sensing circuit and, consequently, not followed by a ventricular spike. With the third P wave, the whole cycle starts all over again, provided that this P wave comes before the end of the programmed V-A interval (as noted earlier VAIp = LRI − AVIp).

When the intrinsic atrial rate augments from very low to high (e.g. with a continuously increasing work load), three different responses will successively arise, as shown in Figure 26. Very low atrial rates will result in pacing at the lower rate limit and between the lower and the upper rate limits, the ventricle will track the atrium. Finally, above the upper rate limit, a 2 : 1 block response will occur and the ventricle will only be paced every other spontaneous atrial event.

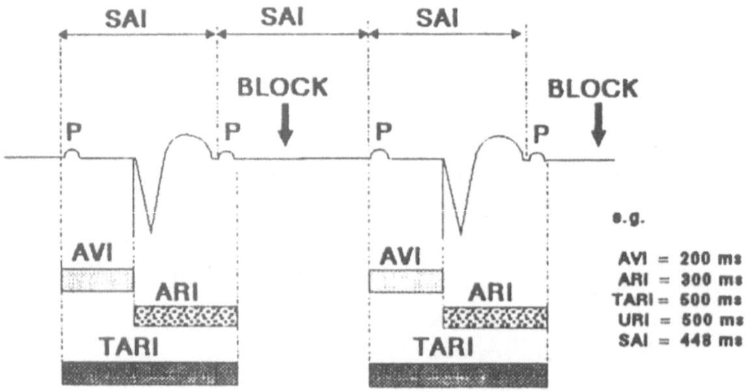

Figure 25. ECG representation of 2:1 block response. Example of 2:1 block response of a DDD pacemaker with TARI = URI and when SAI < URI (or SAR > URL). SAI = spontaneous atrial interval (ms); SAR = spontaneous atrial rate (bpm); TARI = total atrial refractory interval (ms) = AVI + ARI; URI = upper rate interval (ms); URL = upper rate limit (bpm).

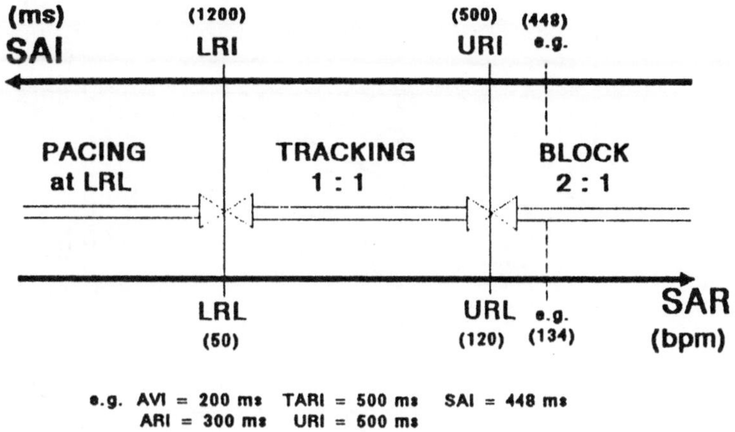

Figure 26. Different pacing responses with the SAR varying from below LRL to beyond URL (or with SAI ranging from larger than LRI to smaller than URI) in patients with total A-V block. Example of different responses of a DDD pacemaker with TARI = URI and same settings as in the ECG of Figure 25.

The diagram of Figure 27 is simply another representation of the same data: it shows the ventricular pacing rate (VPR) as a function of the spontaneous atrial rate (SAR) in patients with total A-V block. Remember that for this case, the upper rate limit (URL) equals the blocking onset rate (BOR) and note the abrupt halving of the ventricular pacing rate when the intrinsic rate reaches this limit. In some patients, the sudden fall of ventricular pacing rate may lead to symptoms of low cardiac output and, therefore, some DDD devices are equipped with a special 'fall back' algorithm.

Pseudo-Wenckebach response:
As already mentioned, a pseudo-Wenckebach behavior may occur when the programmed total atrial refractory interval is shorter than the programmed upper rate interval (TARI < URI or BOR > URL). Obviously, such a pseudo-Wenckebach cycle is only possible when the atrial rate (SAR) exceeds the upper rate limit (URL), but stays below the blocking onset rate (BOR).

An illustration of a Wenckebach cycle can be seen in Figure 28. The first P wave (P1) starts an A-V interval at the end of which a ventricular spike is launched. This ventricular stimulus not only starts a postventricular atrial refractory interval (ARI), but also a new upper rate interval (URI). The second P wave (P2), falling after the atrial refractory interval, is sensed and thus activates an A-V interval. Since this new A-V interval terminates before the end of the upper rate interval, the release of the next ventricular stimulus

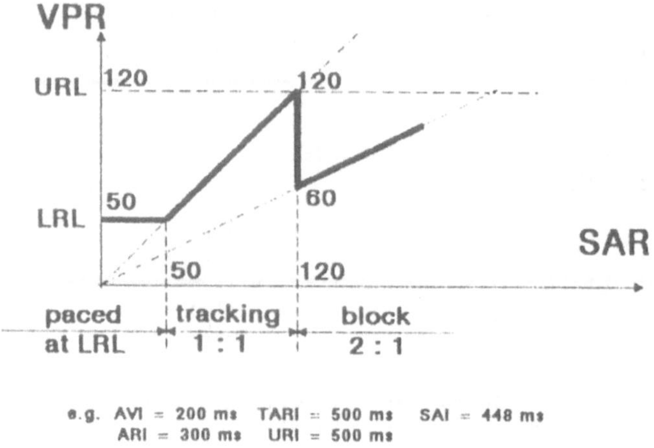

Figure 27. Diagram showing the ventricular pacing rate (VPR) as a function of the spontaneous atrial rate (SAR) in patients with total A-V block. Example of different responses of a DDD pacemaker with TARI = URI and same settings as in Figures 25 and 26.

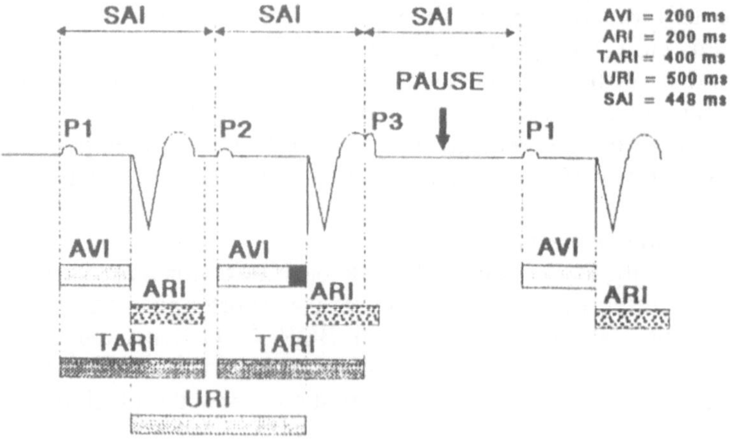

Figure 28. ECG representation of pseudo-Wenckebach response of DDD pacemaker. Example of pseudo-Wenckebach response with TARI < URI and when TARI < SAI < URI (or BOR > SAR > URL). SAI = spontaneous atrial interval (ms); SAR = spontaneous atrial rate (bpm); TARI = total atrial refractory interval (ms) = AVI + ARI; URI = upper rate interval (ms); URL = upper rate limit (bpm); BOR = blocking-onset rate (bpm).

is postponed until this upper rate interval (URI) is completed. Apparently, the A-V interval is lengthened and the prolongation is called the atrioventricular extension period (AVE). The third P wave (P3), falling within the atrial refractory interval (ARI), is obviously not sensed and, hence, not followed by a ventricular stimulus; this results in a pause. The next P wave will start the whole cycle all over again, provided that the atrial rate remains stable but faster than the programmed upper rate limit.

A pseudo-Wenckebach cycle as generated by a DDD pacemaker is characterized by a progressive prolongation between the sensed P wave and the successive ventricular spike until a P wave is not detected, leading to a pause. The resulting atrioventricular extension period (AVE) may range from nil to the so-called Wenckebach interval (WI). This Wenckebach interval or maximum value of the extension is defined by the difference between the upper rate interval and the total atrial refractory interval (Figure 29):

$$\text{WI} = \text{URI} - \text{TARI}$$

Before the pause of an artificial Wenckebach cycle of a DDD pacemaker, the interval between two consecutive paced ventricular events remains constant and equals the upper rate interal (URI). Since this is different from the natural-occurring AV block of the Wenckebach type, the phenomenon with the DDD pacemaker is characterized as 'artifical' or 'pseudo'.

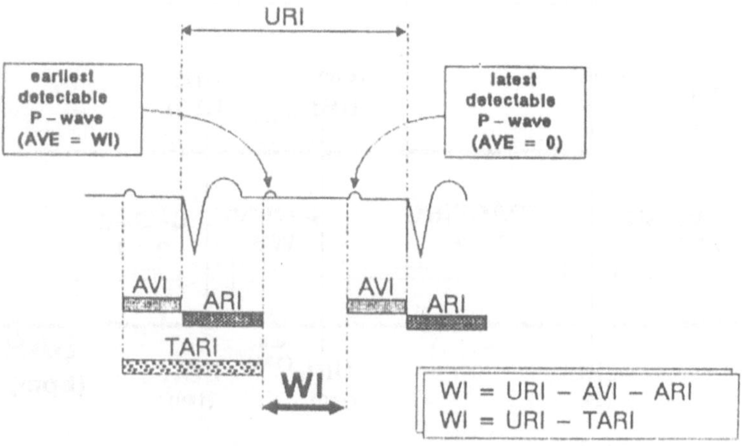

Figure 29. Definition of Wenckebach interval (WI). AVI = programmed A-V interval (ms); ARI = postventricular atrial refractory interval (ms); TARI = total atrial refractory interval (ms) = AVI + ARI; AVE = atrioventricular extension period (ms). Note that the earliest detectable P wave and the latest detectable P wave are superimposed upon the same figure but cannot occur simultaneously.

The ventricular pacing behavior in the absence of natural A-V conduction is schematically depicted in Figure 30, while the ventricular pacing rate (VPR) as a function of the spontaneous atrial rate (SAR) can be found in the diagram of Figure 31. When proceeding from low to high spontaneous atrial rates, several responses will turn up:

* pacing at the lower rate limit (LRL) as long as the intrinsic atrial rate (SAR) remains below this limit,
* 1 : 1 tracking or atrial synchronous ventricular pacing when the sinus rhythm is higher than the lower rate limit (LRL) but lower than the upper rate limit (URL),
* when the spontaneous atrial rate reaches the programmed upper rate limit (URL), the pacing will convert to a pseudo-Wenckebach response (note that although indicated by a horizontal line in Figure 31, the average ventricular pacing rate will be a little lower than the programmed upper rate, due to the pauses!),
* 2:1 block or 2:1 atrial syonchronous ventricular pacing will start if the intrinsic atrial rate (SAR) exceeds the blocking onset rate, i.e. when the spontaneous atrial interval becomes shorter than the total atrial refractory interval (TARI).

Finally, it may be noted that the Wenckebach response has the benefit of smoothing the transition from 1 : 1 tracking to block response. Comparing Figures 27 and 31, it becomes clear that the blocking is postponed to higher

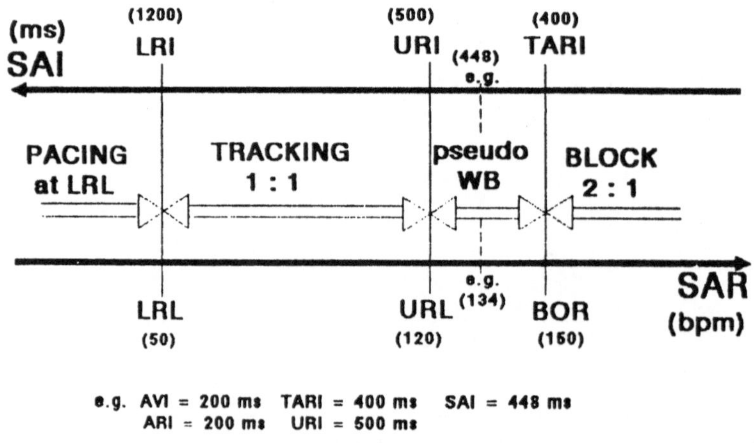

Figure 30. Different pacing responses with the SAR varying from below LRL to beyond BOR (or with SAI ranging from larger than LRI to smaller than TARI) in patients with total A-V block. Example of different responses of a DDD pacemaker with TARI < URI and same settings as in the ECG of Figure 28.

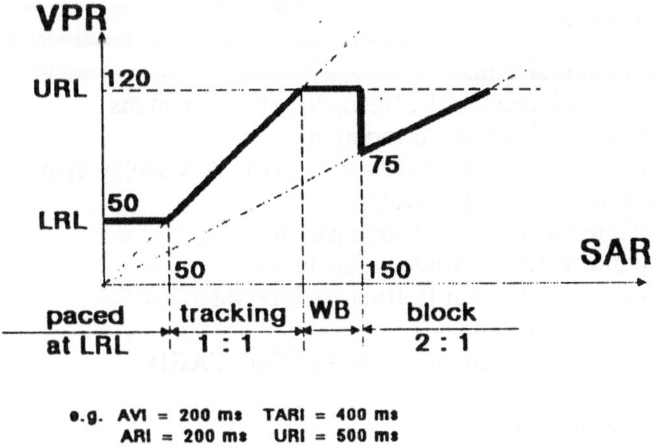

Figure 31. Diagram showing the ventricular pacing rate (VPR) as a function of the spontaneous atrial rate (SAR) in patients with total A-V block. Example of different responses of a DDD pacemaker with TARI < URI and same settings as in Figures 28 and 30.

atrial rates and that the abrupt rate drop is smaller when the programmed values of the pacemaker parameters are properly chosen.

CONCLUSION

During the last decennium, the dual chamber DDD pacing has grown and matured substantially. The initial rather large devices are being succeeded by smaller, longer-lasting pulse generators and most of these pacemakers feature extensive programmability. This progress is, in good measure, due to the amazing evolution of the micro-electronic technology, together with improvements in atrial leads and in lithium batteries.

Today's DDD stimulators truly represent the result of excellent cooperation between engineers and physicians. Rapid advancement of technology and concepts requires constant education to stay abreast and, although even the most sophisticated state-of-the-art devices consist of simple modules, a combined effort by physicians and engineers remains necessary in order to use these 'building-blocks' properly.

USED ABBREVIATIONS

Abbrev.	*Explanation*	*Page*
AI	Alert Interval in ms	49
ARI	(post-ventricular) Atrial Refractory Interval in ms	62
aRP	Absolute Refractory Period in ms	49
AVI	A-V Interval; Atrioventricular Interval; A-V Delay in ms	59
AVIp	Programmed A-V Interval in ms	60
AVIpa	A-V Interval after atrial pacing in ms	59
AVIse	A-V Interval after atrial sensing in ms	59
aVRI	Absolute Ventricular Refractory Interval in ms	62
BI	Basic Interval in ms	44
BOR	Blocking Onset Rate in bpm (= 60000: TARI)	72; 73
bpm	Beats per minute	44 e.or
BR	Basic Rate in bpm	44
EI	Escape Interval in ms	45; 56
ER	Escape Rate in bpm	45; 57
ICHD	Inter-Society Commission for Heart Disease Resources	41
LRI	Lower Rate Interval in ms	56
LRL	Lower Rate Limit in bpm	56
ms	millisecond	44 e.or
mV	millivolt	47 e.or
PVARP	Post-Ventricular Atrial Refractory Interval (= ARI) in ms	62
PVC	Premature Ventricular Contraction	48 e.or
RP	Refractory Period in ms	48
rRP	Relative Refractory Period/Noise Sampling Period in ms	49
rVRI	Relative Ventr. Refrac. Int.; Noise Sampling Interval in ms	62
SAI	Spontaneous Atrial Interval in ms	70
SAR	Spontaneous Atrial Rate in bpm	71
TARI	Total Atrial Refractory Interval in ms (TARI = AVI + ARI)	63
URI	Upper Rate Interval in ms	57
URL	Upper Rate Limit in bpm	59; 72
VAI	V-A Interval; Ventriculo-atrial Interval in ms	60
VAIp	Programmed V-A Interval in ms	60
VBI	Ventricular Blanking Interval in ms	66
VPI	Ventricular Pacing Interval in ms	71
VPR	Ventricular Pacing Rate in bpm	71
VRI	Ventricular Refractory Interval in ms	61
VTL	Ventricular Tracking Limit (= URL) in bpm	72

PACEMAKER MODES

Abbrev.	*Explanation*	*Page*
AAI	Atrial pacing; Atrial sensing; Inhibited mode	51
DDD	Dual chamber pacing; Dual chamber sensing; Inh. & Trig. mode	41; 52; 53
DDDR	Dual pac.; Dual sens.; Inh. & Trig.; Rate responsive	42
DVI	Dual chamber pacing; Ventricular sensing; Inhibited mode	41; 52
VAT	Ventricular pacing; Atrial Sensing; Triggered mode	41; 51
VDD	Ventricular pacing; Dual chamber sensing; Inh. & Trig. mode	41; 52
VVI	Ventricular pacing; Ventricular sensing; Inhibited mode	41; 42; 44
VVIR	Ventr. pac.; Ventr. sens.; Inhib. mode; Rate responsive	42

5. A practical guide to the interpretation of DDD pacing electrocardiograms

ROLAND STROOBANDT

The interpretation of DDD pacing electrocardiograms may be complex and requires thorough knowledge of the operational concepts. In the DDD mode, the pulse generator operates with sensing and stimulation at both the atrial and ventricular levels depending on the presence or absence of spontaneous cardiac activity in relation to the programmed intervals. It is impossible to give a complete survey of all electrocardiographic appearances. This chapter will rather provide a framework for a systematic approach to the interpretation of DDD pacemaker electrocardiograms. The analysis of pacemaker electrocardiograms may be facilitated by answering to five key questions:

1. What is the underlying heart rhythm?
2. Are there pacemaker stimuli visible?
3. Do the stimuli result in capture?
4. Is there proper sensing of P waves and QRS complexes?
5. Is the atrial and/or ventricular rate in accordance to the programmed rate? (see Table 1)

UNDERLYING HEART RHYTHM

A pacemaker electrocardiogram has to be approached at first as a nonpaced electrocardiogram. One should start by screening the electrocardiogram for any spontaneous cardiac activity as well at the atrial as at the ventricular level at any portion of the electrocardiogram during which the heart is unpaced (Table 2). Moreover one should try to determine whether there exists a relationship between the native atrial and ventricular activity and whether the QRS complexes are small or broad. When no sinus P waves can be detected, every attempt should be made to document the patient's underlying rhythm. If the pacemaker is functioning continuously and when there is no evidence of the intrinsic rhythm, the pacemaker should be reprogrammed to a lower rate to unmask the patient's rhythm.

E. Andries, P. Brugada & R. Stroobandt (eds.), How to face 'the faces' of cardiac pacing, 83—117.

Table 1. Key questions.

— Underlying rhythm
— Pacing
— Capture
— Sensing
— Rate

Table 2. Underlying rhythm.

Atrial:	— Sinus rhythm?
	— Atrial flutter/fibrillation?
Ventricular:	— Morphology of QRS complex?
AV relation:	— 1/1 conduction?
	— AV block/dissociation?

2. PACING

Following the first screening of the electrocardiogram for the patient's underlying rhythm, the ECG should be scrutinized for the presence of atrial and ventricular stimuli (Table 3). Complete absence of stimuli may occur during normal operation of a DDD pacemaker whenever the pacemaker is completely inhibited by the spontaneous heart rhythm of the patient that is faster than the programmed lower rate. To assure that the pacemaker is capable of emitting stimuli, it may be necessary to perform carotid sinus massage (Figure 1) to slow the heart rhythm of the patient below the preset lower rate or to apply a magnet over the pacemaker which will result, in most circumstances, in asynchronous (DOO) pacing in both the atrial and ventricular chamber (Figure 2).

Examination of the configuration of the *pacemaker artifact* itself may reveal whether it concerns a unipolar or a bipolar pacemaker: unipolar spikes are much larger than bipolar ones because of the increased distance between the poles results in a bigger electric dipole (Table 4). Significant differences in the amplitude of the stimulus artifact may be due to several causes [1]. Programming a lower output will result in a smaller artifact. Slight changes in amplitude of the pacemaker artifact may occur, especially in V1, and are simply due to respiration. It might also indicate the presence of an insulation defect or a partial wire fracture. Likewise, oversaturation of the ECG amplifier or altered filter settings of the ECG recorder (Figure 3) may affect the configuration of the pacemaker artifact.

Pacemaker stimuli may be absent even after magnet application or carotid sinus massage (Table 5). *No output* may be caused by battery depletion, lead

Table 3. Pacemaker stimuli: present *or* absent.

Present:	— Atrial, ventricular *or* both?
Absent:	— Normal function (PM inhibited) — Apply magnet — Carotid sinus massage

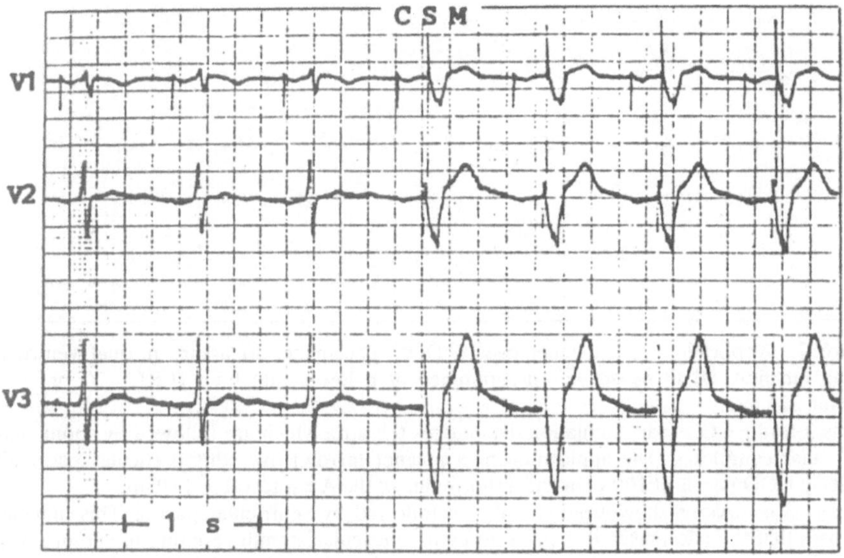

Figure 1. Carotid sinus massage to disclose normal pacemaker function. DDD pacemaker programmed at lower rate limit (LRL): 70 ppm; AV interval: 225 ms.
The first 3 complexes show atrial stimulation followed by spontaneous narrow QRS complexes. From the 4th beat on carotid sinus massage is performed which prolongs the AV conduction time and ventricular stimulation becomes visible.

fracture or to a loose set screw. A sudden loss of pacemaker output may occur soon after the replacement of a unipolar pacemaker by a smaller unit, so that air get entrapped and muscle contact is lost. A sterile needle puncture of the pacemaker pocket may be needed to restore pacing.

3. CAPTURE

Documentation of atrial and ventricular *capture* is critical and pivotal to the issue of dual chamber pacing. Particularly, the recognition of atrial capture is

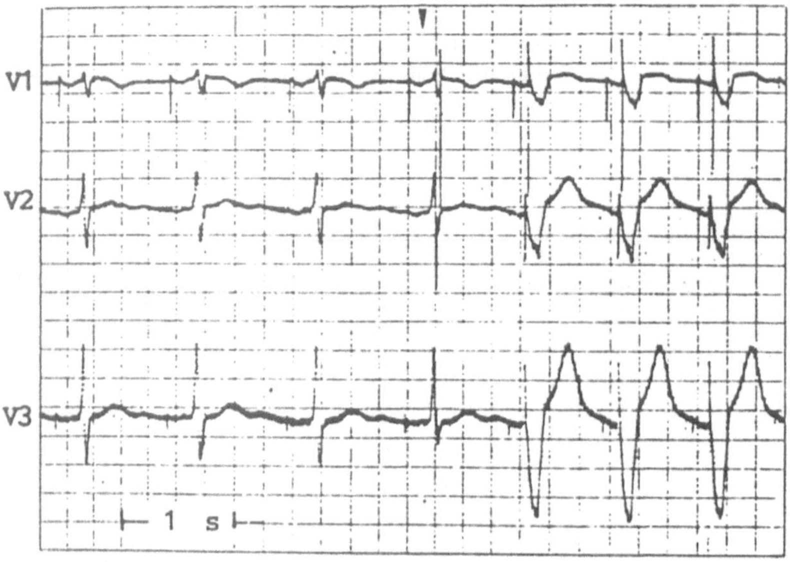

Figure 2. Application of a magnet over a DDD pacemaker to assure normal ventricular pacing function. DDD pacemaker programmed at a lower rate limit (LRL): 70 ppm; AV interval: 225 ms.
In this 3-lead ECG, atrial stimulation depolarizes the atria which are followed by spontaneous ventricular complexes. The application of a magnet (arrow) over this particular pacemaker results in DOO pacing at 100 ppm and a shortening of the AV interval to 110 ms.
The last 3 complexes show atrial stimulation followed by ventricular pacing. This maneuver discloses that the pacemaker is capable to emit ventricular stimuli resulting in capture of the ventricle in a patient in whom the ventricular pacemaker channel was inhibited by spontaneous ventricular complexes.

Table 4. Pacemaker stimuli: configuration of PM artifact.

— Unipolar artifact > bipolar artifact
— Same configuration of stimulus artifact?
  — Lower output?
  — Respiration?
  — Defective electrode insulation?
  — Partial wire fracture?
  — Saturation of ECG amplifier?
  — Filter settings of ECG recorder?

mandatory as the P wave may be obscured by the large unipolar stimulus overshoot (Table 6). To find out whether there is capture, one should screen all leads of the ECG tracing. Even then it is not uncommon that the 12-lead ECG fails to permit any conclusion (Figure 4). The use of a double standar-

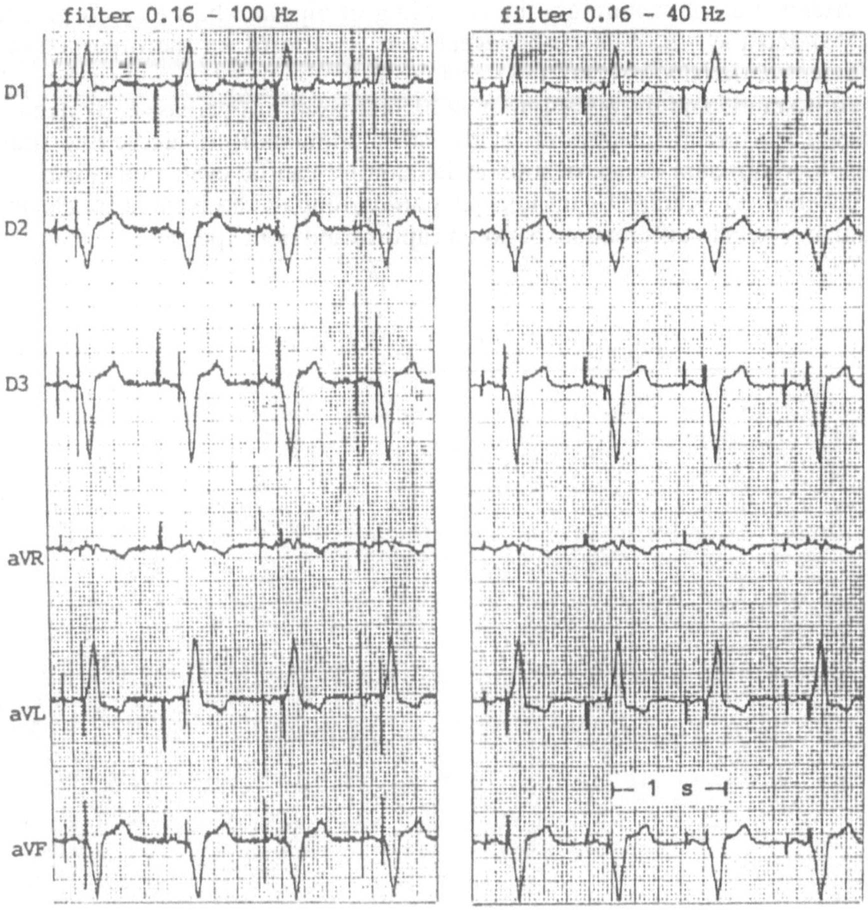

Figure 3. Effect of filter settings on pacemaker artifact. Unipolar DDD pacemaker programmed at a lower rate limit (LRL): of 70 ppm; AV interval: 180 ms.
Left panel: Filter settings range from 0.16 to 100 Hz; both atrial and ventricular pacemaker stimuli are clearly visible. The variable amplitude of the spikes is due to digital conversion of the signal.
Right panel: By cutting off the higher frequencies beyond 40 Hz, the amplitude of the pacemaker artifacts is largely reduced.

Table 5. Pacemaker stimuli absent.

Causes of no output:	— Power source failure
	— Lead fracture
	— Loose set screw
	— Air entrapment in pacemaker pocket

dization may disclose otherwise invisible P waves (Figure 5). Other means of ascertaining atrial capture are the recording of bipolar chest leads (e.g. V3R — V3) and the use of an oesophageal lead (Figure 6). In the presence of a sufficient underlying rhythm, the pacemaker can be programmed to the AAI or AOO mode and the atrial rate may be increased or decreased. By demonstrating a constant relationship of the atrial stimulus to a succeeding spontaneous ventricular depolarization atrial capture can be evaluated (Figure 7). In pacemakers equipped with telemetry, recording of the intra-atrial electrogram may help to differentiate between capture and noncapture.

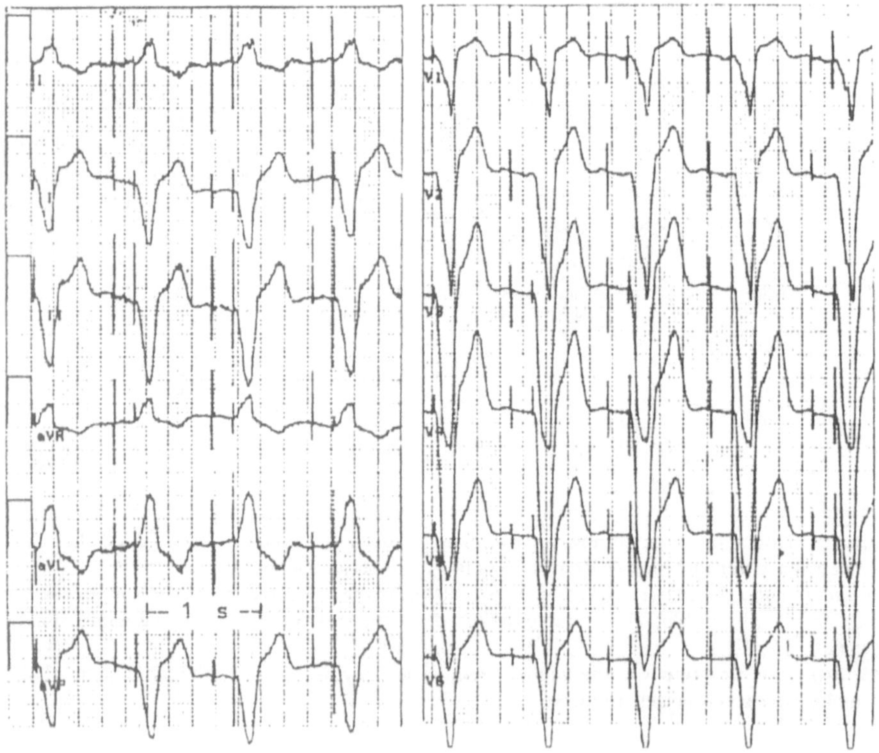

Figure 4. Atrial capture? Unipolar DDD pacemaker programmed at a lower rate limit (LRL): 70 ppm; AV interval: 180 ms; Atrial and ventricular output 5 V; Atrial and ventricular pulse duration 0.5 ms.
From this 12-lead ECG it is not possible to determine whether the atrial stimulus is followed by atrial capture and other measures have to be taken to know whether there is atrial capture.

Table 6. Atrial capture.

Successful atrial capture?	
	— 12-lead ECG
	— ECG recording at double standardization
	— Bipolar chest lead, e.g. V3r- V3
	— Decrease *or* increase atrial output
	— Rate decrease/increase in AAI *or* AOO mode
	— M-mode, 2D echocardiography
	— TRansmitral doppler flow
	— Esophageal lead
	— Telemetry

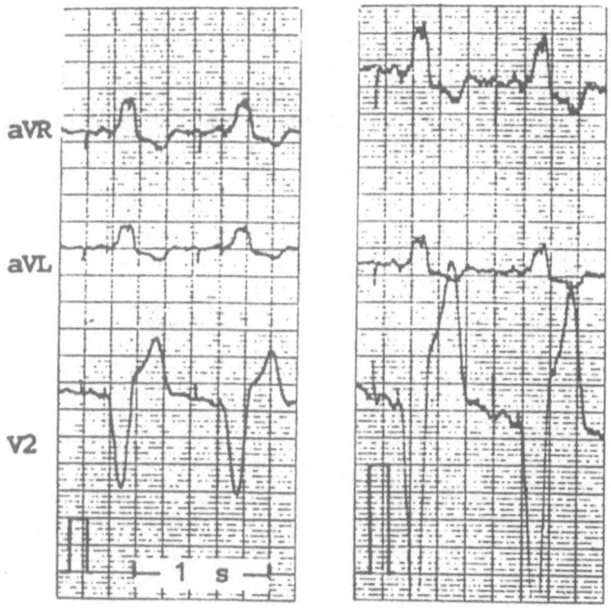

Figure 5. Use of double standardization to disclose atrial capture. DDD pacemaker programmed at a lower rate limit (LRL): 70 ppm; AV interval: 180 ms.
Left panel: ECG taken at normal standardization (1 mV = 10 mm).
Right panel: By using double standardization (1 mV = 20 mm) atrial capture becomes evident in lead V2.

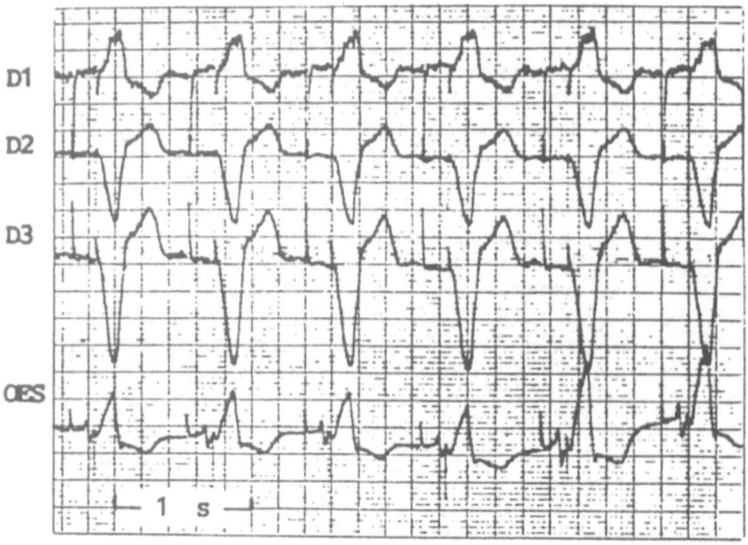

Figure 6. Use of oesophageal lead to demonstrate atrial capture. DDD pacemaker programmed at a lower rate limit (LRL): 70 ppm; AV interval: 180 ms.
Atrial depolarization cannot be evaluated in lead D1-D3. A deflection preceding the QRS complex on the oesophageal lead recording (oes) ensures that there is atrial capture.

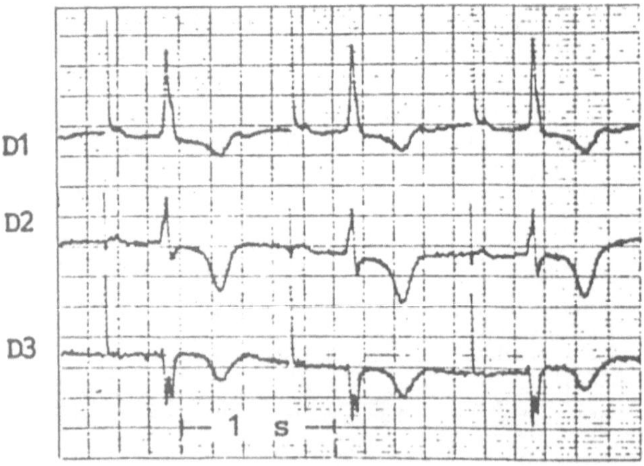

Figure 7. Proof of atrial capture by programming DDD pacemaker in AAI or AOO mode. DDD pacemaker reprogrammed in the AAI mode at 50 ppm.
The atrial stimulus depolarizes the atrium which is followed after a prolonged AV interval by a spontaneously conducted QRS complex, proofing that there is atrial capture. The deep negative T waves are 'postpacing T waves' which may appear after a prolonged period of ventricular pacing.

M-mode, 2D echocardiography or recording of an A wave in the ventricular doppler inflow signal may give definite proof of atrial capture (Figure 8).

Other signs may help to detect a loss of atrial capture (Table 7). At physical examination, a varying intensity of the first heart sound suggests AV dissociation. The presence of regular cannon waves in the jugular venous pulse suggests constant retrograde VA conduction and, consequently, reflects lack of atrial capture. Furthermore, notching of the ST-segment or the presence of a deformity of the T wave on the ECG may indicate atrial noncapture (Figure 9).

Table 7. Atrial noncapture.

Search for retrograde atrial activation:	
	— Varying intensity of 1st heart sound
	— Cannon waves in jugular venous pulse
	— Notching of the ST segment
	— Deformity of the T wave

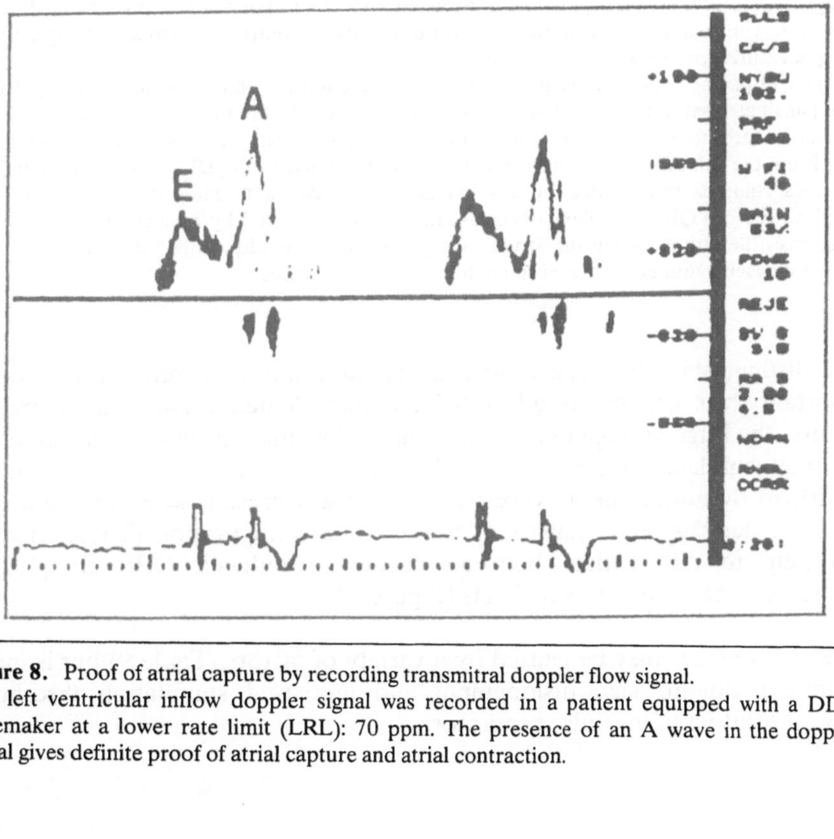

Figure 8. Proof of atrial capture by recording transmitral doppler flow signal.
The left ventricular inflow doppler signal was recorded in a patient equipped with a DDD pacemaker at a lower rate limit (LRL): 70 ppm. The presence of an A wave in the doppler signal gives definite proof of atrial capture and atrial contraction.

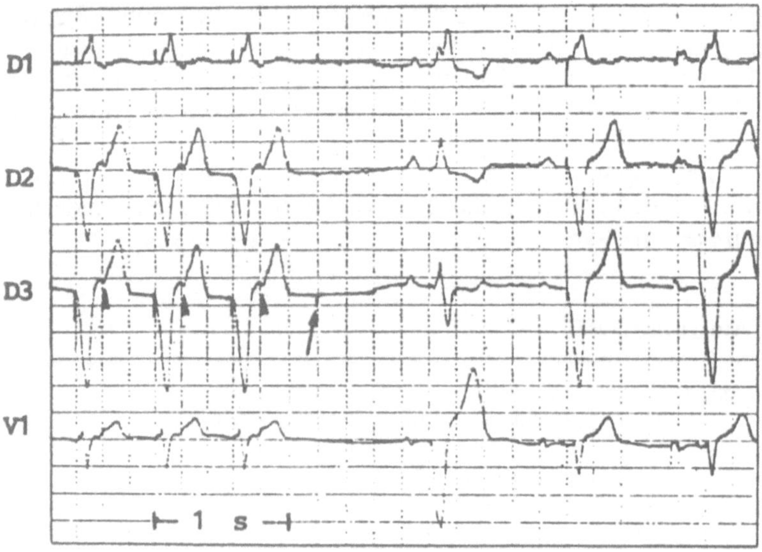

Figure 9. DDD pacemaker temporarily programmed in the VVI mode to demonstrate retrograde VA conduction during loss of atrial capture. The pacemaker is programmed in the VVI mode 100 ppm (first 4 complexes) and the ventricular output is decreased progressively during measurement of the ventricular threshold.
During VVI pacing, there is retrograde activation of the atrium which is visible as a negative P wave (small arrows) in the ST segment. The output of the 4th stimulus is below the ventricular stimulation threshold, resulting in a loss of capture (arrow). The 4th QRS complex is a sinus beat. From the 5th beat on, the pacemaker is reprogrammed to the DDD mode 70 ppm. The 5th QRS complex is preceded by a spontaneous P wave and followed by a paced QRS complex. The 6th QRS complex is pacing-induced and preceded by an atrial stimulus which depolarizes the atrium. When the atrium is depolarized antegradely either spontaneously (5th beat) or by pacing (6th beat), there is no notching of the ST segment.

In distinguishing between ventricular capture and noncapture, there is one important trap: i.e. the pseudo-QRS complex. When there is a failure to capture, the large voltage decay curve may be prominent in unipolar systems and may simulate a QRS complex (Figure 10). Ventricular capture can be confirmed by noting the T wave which is always present when the ventricle has been depolarized. Analysis of the T wave configuration (Table 8) may also help to differentiate between pacemaker-induced QRS complexes, fusion beats, and pseudofusion beats (Figure 11).

Failure to capture may be caused by a variety of factors (Table 9) but is most commonly due to lead dislodgment and increased stimulation threshold above the output of the pulse generator.

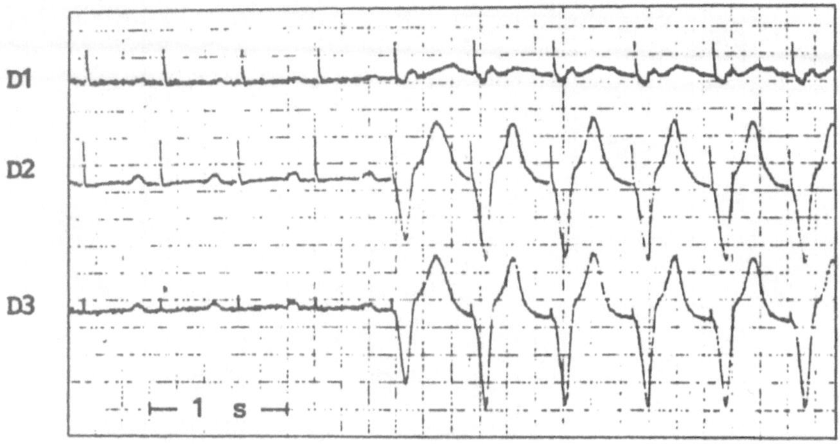

Figure 10. Pseudo-QRS complexes by loss of ventricular capture. DDD pacemaker programmed at a lower rate limit (LRL): 70 ppm; upper rate limit (URL): 150 ppm; AV interval: 200 ms.
Left part: Presence of sinus P waves at a rate of 105 ppm, which are followed after an AV interval of 200 ms by a ventricular stimulus. The stimuli are not followed by a QRS complex nor by a T wave. These are so-called 'pseudo-QRS complexes'. This period of pseudo-pacing results in ventricular asystole.
Right part: Normal DDD pacing is present during the last 6 complexes. The cause of ventricular noncapture is presumably due to dislocation of the ventricular lead as the paced QRS complex is not positive in lead D1. This means that the tip of the lead is not located in the right ventricular apex.

Table 8. Ventricular capture.

Successful ventricular capture.

Distinguish between:
- Pseudo-QRS complexes
  - decay of stimulus artifact
- Pacemaker-induced QRS complexes
- Fusion beats
  - simultaneous depolarization via pacing and normal AV conduction
- Pseudofusion beats
  - spontaneous induced depolarization together with PM spike

Look at T wave configuration

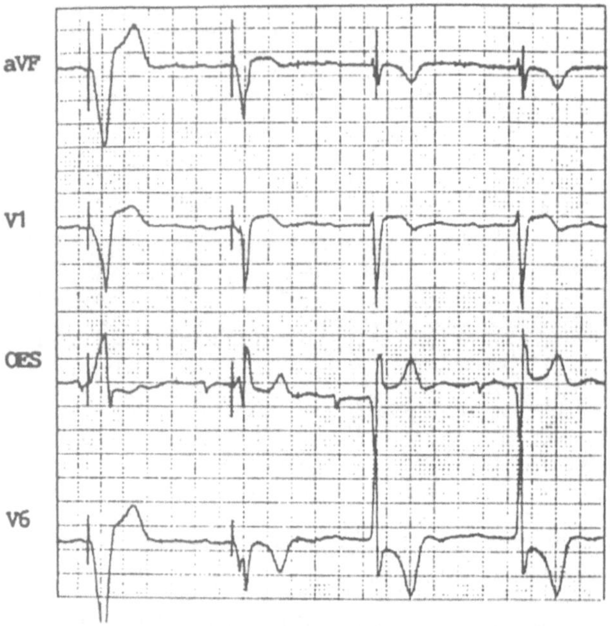

Figure 11. Pacemaker-induced QRS complexes, fusion beats and pseudofusion beats. DDD pacemaker reprogrammed in the VVI mode 50 ppm to demonstrate different degrees of fusion. Lead aVF, V1, V6 are recorded simultaneously with an oesophageal lead.
The 1st QRS complex is a pacemaker-induced QRS complex; the 2nd QRS complex is a fusion beat between a spontaneous QRS complex and a pacemaker-induced QRS complex; the 3rd and 4th QRS complex are pseudofusion beats as the pacemaker spike coincides with the spontaneous QRS complex. The degree of fusion can be evaluated at best by looking at the configuration of the T wave in lead aVF and V6.

Table 9. Ventricular noncapture.

Causes:	— Increased stimulation threshold — Low ventricular output setting — Lead/connector problem

Table 10. Paced QRS complex.

Configuration (RV apex pacing)	— Positive: leads I, aVR, aVL — Negative: leads II, III, aVF — LBBB: precordial leads

During pacing from the right ventricular apex, the *paced QRS complex* should be positive in lead I, AVR & AVL and negative in the inferior leads (II, III, & AVF) and show a complete left bundle branch block pattern in the precordial leads (Table 10). The mean electrical axis is deviated superiorly and usually to the left because the depolarization wave front travels from the apex to the base and away from the inferior leads (Figures 12 and 13).

4. SENSING

The next key question is whether there is adequate sensing of atrial and ventricular events. The occurrence of pacemaker stimuli at an unexpected time suggests malsensing. There are basically two types of sensing problems: undersensing and oversensing (Table 11).

Undersensing occurs when there is failure to sense intrinsic cardiac depolarizations (Figure 14). The most common cause of undersensing is poor signal

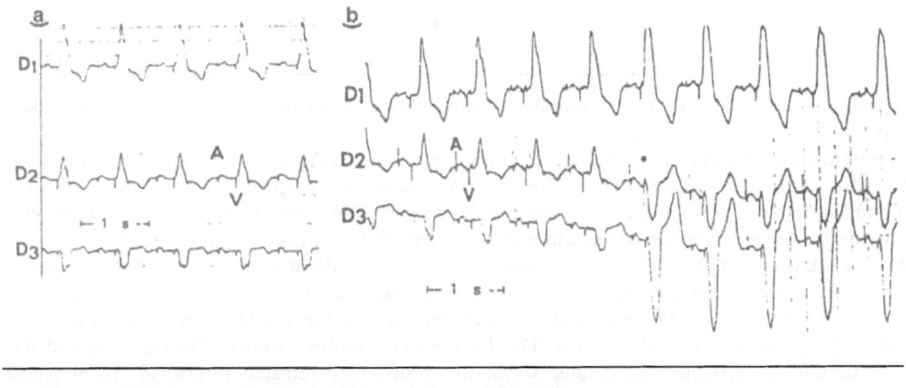

Figure 12. Importance of paced QRS complex configuration.
(a) pacemaker programmed at a lower rate limit (LRL): 70 ppm; AV interval: 250 ms. The pacemaker seems to function normally at first glance; the atrial stimulus (A) is followed by an atrial depolarization and the ventricular stimulus (V) by a broad QRS complex. However, as the heart is paced from the right ventricle, the QRS complex in D2 should be negative. Therefore, this tracing may be suspect for potential lead dislodgment.
(b) Same patient as in ECG 12(a). The DDD pacemaker was reprogrammed to a higher lower rate limit (LRL): 75 ppm and the AV interval shortened to 175 ms. *Left part:* The first 4 complexes show atrial stimulation (A) followed by a ventricular stimulus (V) at 175 ms. The ventricular stimulus does not capture the ventricle. The broad QRS complexes are due to spontaneous ventricular depolarization 250 ms after the atrial stimulus. This proofs that the broad QRS complexes of ECG 12(a) were spontaneous complexes and not pacemaker-induced. *Right part:* During deep inspiration (asterisk), normal DDD pacing can be observed. There is capture of the ventricle 175 ms after the atrial stimulus and the QRS complexes in lead D2 and D3 are negative.

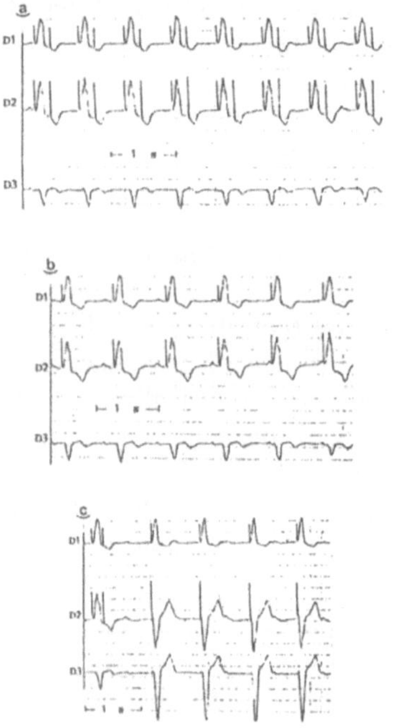

Figure 13. Use of paced QRS complex configuration for location of dislodged pacemaker lead.
(a) Application of a magnet over a DDD pulse generator results in DOO pacing at a magnetic rate of 83 ppm and a AV interval: 250 ms. Two spikes can be observed, one preceding and one following the QRS complex. The spike preceding the QRS complex captures the ventricle; however the QRS complex is positive in lead D2 during right ventricular pacing.
(b) The DDD pacemaker is reprogrammed to the AAO mode at a rate of 70 ppm. The atrial stimulus now paces the ventricle. However the paced QRS complex is positive in lead D2.
(c) First QRS complex: The pacemaker is programmed in the DDD mode; lower rate limit (LRL): 70 ppm; AV interval: 200 ms. The 1st (atrial) stimulus captures the ventricle and the 2nd stimulus is ineffective as it falls within the ventricular refractory interval. Second and following QRS complexes: The pacemaker is reprogrammed to the VVI mode. The ventricular stimulus also captures the ventricle but the QRS complex is now negative in lead D2.
The different tracings (a—c) demonstrate that both the atrial and the ventricular lead are located in the ventricle. The ventricular lead is situated at the right ventricular apex. The atrial lead is dislodged into the right ventricle and is situated more proximal towards the base of the ventricle. The atrial stimulus captures the ventricle before the ventricular stimulus as it is delivered 200 ms earlier.

Table 11. Sensing problems.

– *Undersensing:*	failure to sense intrinsic cardiac depolarizations
– *Oversensing:*	pacemaker senses signals other than R or P waves

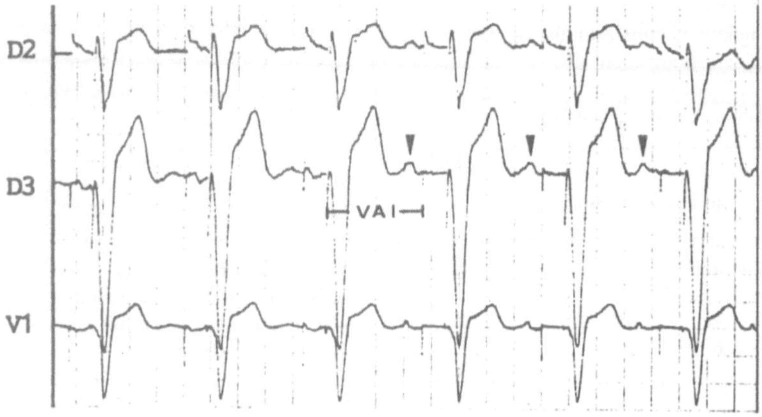

Figure 14. Atrial undersensing: DDD pacemaker programmed at a lower rate limit (LRL): 70 ppm; AV interval: 175 ms; atrial sensitivity 3 mV.
There is normal lower rate DDD pacing (first 3 complexes). The 4th to 6th P wave (arrows) are not sensed. The 4th and subsequent atrial stimuli occur at a normal ventriculoatrial interval (VAI) but do not capture the atria because of the biological refractoriness of the atria.

detection because of nonoptimal placement or dislodgment of the lead. Far less frequently, it is due to a lead or connector problem and it is rather seldom due to a failure of the pacemaker sensing circuit (Table 12).

Oversensing is present when the pacemaker senses signals other than P or R waves. This resets the pacemaker resulting in an interval that is greater than the programmed cycle length. Oversensing may be caused by a variety of factors (Table 13). The heart itself may be the source as there may be oversensing of the QRS complex or the T wave (Case study 1: Figure 26). In unipolar systems, sensing of myopotentials is by far the most common cause of oversensing. In these systems, the pacemaker can comprises the indifferent electrode. Contraction of the muscle in contact with the pacemaker may generate sufficiently large signals to inhibit the pacemaker (Figure 15). When specifically looking for pacemaker inhibition by myopotentials during follow-up, the electrocardiogram may be completely distorted and difficult to interpret. To overcome these problems one of the arm electrodes can be placed on the forehead which eliminates a lot of noise and makes the pacemaker spike more visible.

The pacemaker itself may also be the source of oversensing as the ventricular channel of the pacemaker may sense afterpotentials from the output of the atrial channel and be inhibited (Table 14). This is known as crosstalk or self-

Table 12. Causes of undersensing.

- Poor signal detection (amplitude/slew rate)
- Electrode dislodgement
- Lead / connector problem
- Pacemaker sensing circuit failure

Table 13. Cause of oversening.

— Myopotentials	—Crosstalk/far field signals
— T wave sensing	— Electromagnetic interference
— QRS sensing	— Pacemaker sensing circuit failure
— Afterpotentials	— Concealed extrasystoles

Table 14. Crosstalk.

Causes:	inappropriate programming of — High atrial output — Short blanking period — High ventricular sensitivity

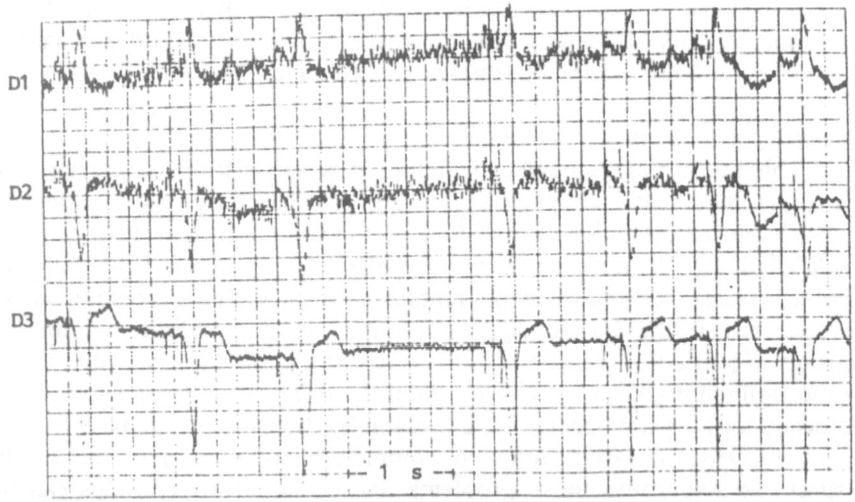

Figure 15. Pacemaker output inhibition by sensing of myopotentials. DDD pacemaker programmed at a lower rate limit (LRL): 75 ppm; AV interval: 165 ms; atrial sensitivity: 1 mV; ventricular sensitivity: 2 mV.

ECG leads D1, D2, D3 are recorded during contraction of the muscles of the forearm. Myopotentials are clearly visible in lead D1 and D2. Intermittent sensing of myopotentials results in resetting of the pacemaker and a slowing of the pacing rate. Reprogramming the atrial and ventricular sensitivity to higher values could solve the problem.

inhibition. The most common cause of crosstalk is an inappropriate programming of a high atrial output, a short ventricular blanking period, and a high ventricular sensitivity (Figure 16). False signals due to a partial lead fracture or loose connection between lead and connector or by intense electromagnetic fields, may also cause oversensing.

For the management of undersensing, the pacemaker has to be reprogrammed to a higher sensitivity (e.g., from 2 to 1 mV). Oversensing may be treated by reprogramming the pacemaker to a lower sensitivity (e.g., from 1 to 2 mV). However, reprogramming a longer postventricular atrial refractory period, a longer blanking period, and reducing the output of the pacemaker, should also be considered for the management of crosstalk (Table 15).

5. RATE

To examine whether the pacemaker is functioning normally, an exact understanding of the various timing cycles of the DDD pacemaker is necessary (see Chapter 4 by A. Sinnaeve). Moreover, the different settings to which the pacemaker is programmed should be known either from the medical record or by interrogating the device.

1. *Lower rate operation*

When atrial pacing is observed during operation in the DDD mode, the pacemaker is always functioning at its lower rate. The lower rate interval (LRI) which is given by dividing 60 000 by the lower rate limit, is the largest interval between two consecutive paced ventricular events or between a sensed spontaneous ventricular event and the subsequent paced event (in the absence of hysteresis). This lower rate interval (LRI) can be subdivided into two portions: the ventriculo-atrial interval and the programmed AV interval (AVIp). The ventriculoatrial interval (VAI) begins with ventricular stimulus or ventricular sensed event and ends at the atrial stimulus (Figure 17).

Measurement of the ventriculoatrial (VAI) may be very helpful in the analysis of the DDD pacemaker electrocardiogram to ascertain normal

Table 15. Malsensing.

Management:	— Undersensing		
		— reprogram *higher* sensitivity, e.g. from 2 mV to 1 mV	
	— Oversensing		
		— reprogram *lower* sensitivity, e.g. from 1 mV to 2 mV	
		— consider reprogramming:	
			— longer ARI
			— longer blanking period
			— lower output

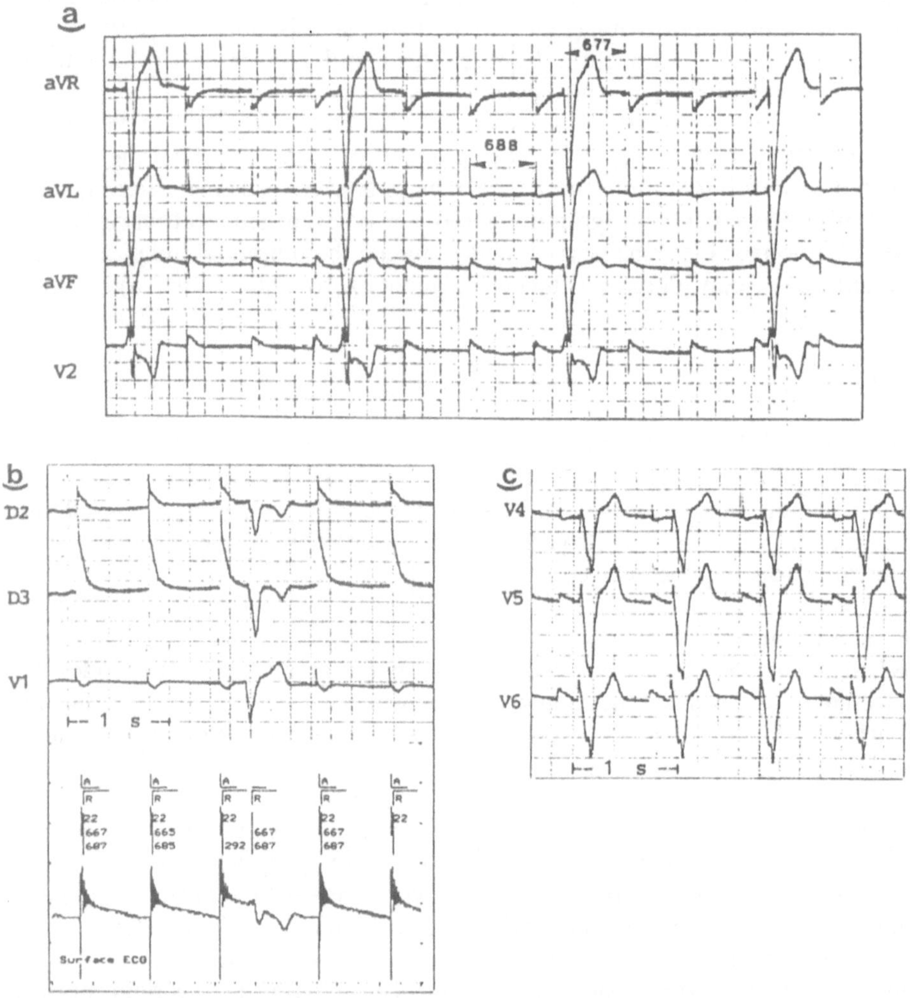

Figure 16. Inhibition of ventricular output by crosstalk: resulting in 'pacemaker acceleration'. DDD pacemaker programmed at a lower rate limit (LRL): 70 ppm, i.e. lower rate interval (LRI) 857 ms; AV interval: 180 ms; atrial output 10 Volt, 1 ms; ventricular sensitivity 0.5 mV; ventricular blanking period 13 ms.

(a) There is an underlying idioventricular heart rhythm at a rate of 25 beats per minute. Although a lower rate limit (LRL) of 70 ppm was programmed, the pacemaker initiates stimuli at a rate of 87 ppm (interstimulus interval 688 ms) i.e. 'pseudoaccleration'. There is appropriate sensing of the ventricular QRS complex as the atrial stimulus is reset by the ventricular event and resumes pacing after a normal ventriculoatrial interval (VAI) of 677 ms; VAI = LRI — AVIp, i.e. in this case: 677 = 857 — 180. No ventricular pacing stimuli are visible on the ECG tracing.

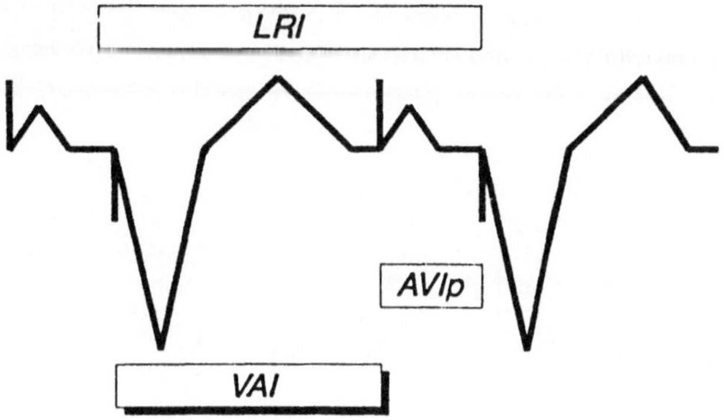

Figure 17. Shematic diagram of the ventriculoatrial (VAI) interval. LRI: lower rate interval; AVIp: programmed atrioventricular interval.

pacemaker operation. The ventriculoatrial interval (VAI) can be calculated by subtracting the programmed AV interval (AVIp) from the lower rate interval:

$$VAI = LRI - AVIp.$$

During normal operation at its basic (lower) rate (Figure 18–19), the ventriculoatrial interval (VAI) has always to be equal to the (LRI − AVIp).

Only when special modes such as hysteresis or 'rate drop at night' are programmed, the ventriculoatrial interval VAI may be greater than (LRI − AVIp). In all other circumstances, a VAI > (LRI − AVIp) is abnormal and might indicate either false inhibition (see Figure 26) or impending battery depletion with a slowing of the basic pacing rate.

A ventriculoatrial interval VAI < (LRI − AVIp) can be observed when the DDD(R) mode or other special modes such as rate smoothing and flywheel have been programmed, otherwise it is caused by a runaway (Figure 20).

(b) The ECG tracing shows a large decay curve following the atrial stimuli in lead D2 and D3 due to the high atrial output setting of 10 Volt, 1 ms. Telemetry discloses sensing of the afterpotential by the ventricular pacemaker channel 22 ms after the delivery of the atrial stimulus. This is interpreted by the pacemaker as a ventricular event resulting in the inhibition of the ventricular channel.

(c) By reprogramming the ventricular blanking period from 13 to 50 ms, the ventricular channel no longer senses the afterpotential of the atrial output pulse and normal DDD pacing is restored at 70 ppm.

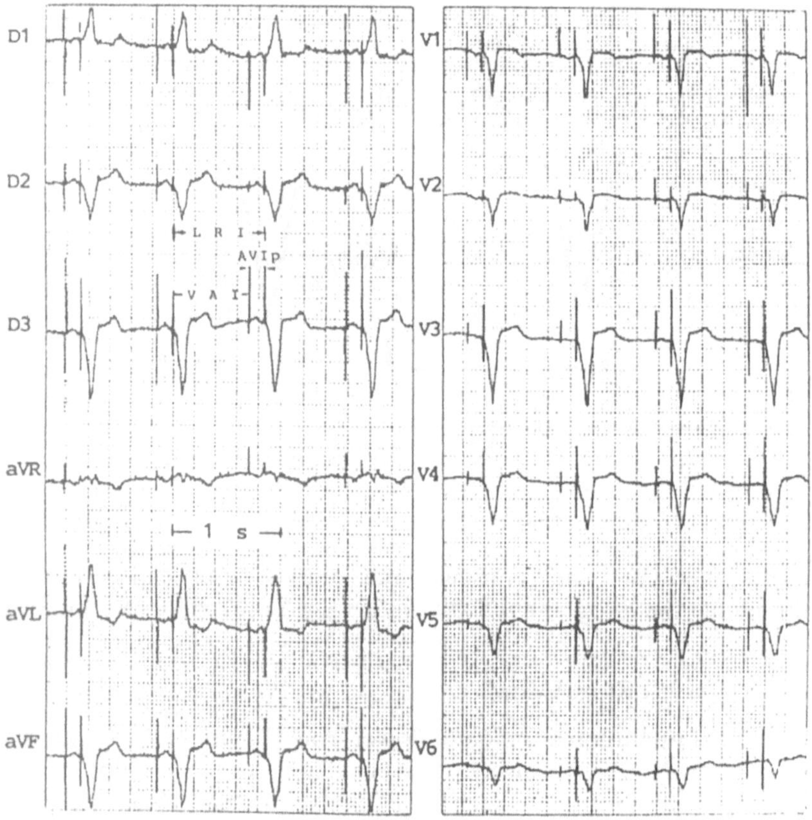

Figure 18. Calculation of the ventriculoatrial interval (VAI) during pacing at the lower rate limit (basic pacing rate). DDD pacing at a lower rate limit (LRL): 70 ppm; AV interval: 150 ms.
The lower rate interval (LRI) can be calculated by dividing 60 000 by 70 which equals 857 ms. The AV interval is programmed (AVIp) at 150 ms. The ventriculoatrial interval (VAI) during lower rate DDD pacing is constant and equals: LRI — AVIp = VAI, i.e. in this case: 857 — 150 = 707 ms.

2. *Atrial tracking*

DDD pacemakers are devices capable of sensing and pacing in both atrium and ventricle and can synchronize atrial and ventricular events. To safeguard against excessively rapid ventricular rates with the occurrence of atrial flutter or fibrillation, a need was perceived to limit the maximum stimulation rate of the ventricular channel. The upper rate limit (URL) is the shortest interval between ventricular paced events or a sensed event followed by a paced event. Whenever the spontaneous atrial rate (SAR) varies between the preset lower rate limit (LRL) and upper rate limit (URL), the ventricular

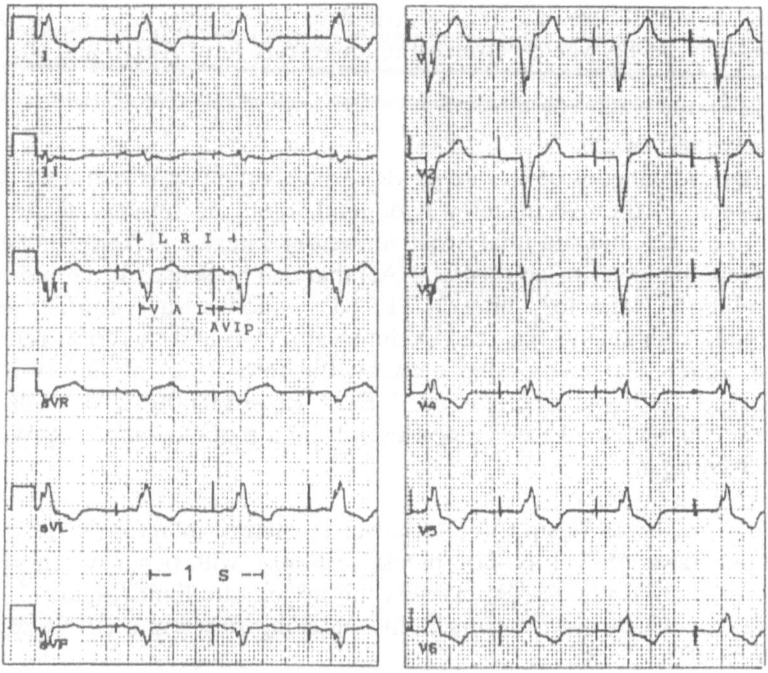

Figure 19. Calculation of the ventriculoatrial interval (VAI) during lower (basic) rate pacing. DDD pacemaker programmed at a lower rate limit (LRL): 70 ppm; lower rate interval (LRI = 60 000: LRL = 857 ms); AV interval 250 ms.
The lower rate interval (LRI) equals 857 ms. The ventriculoatrial interval (VAI) during lower rate pacing is constant and equals: LRI — AVIp = VAI, i.e. in this case 857 — 250 = 607 ms.

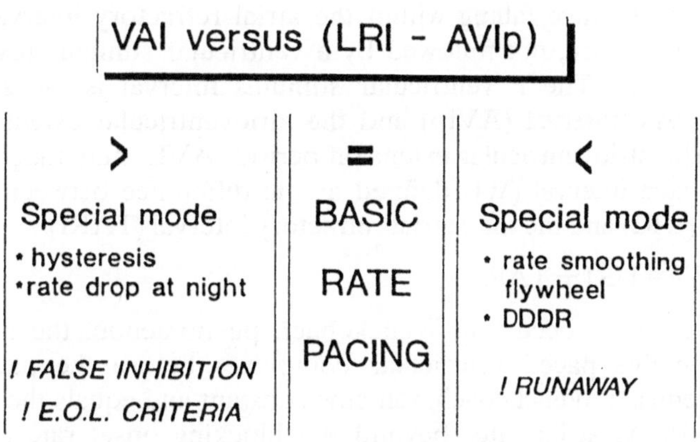

Figure 20. Usefulness of the ventriculoatrial interval (VAI) in the differentiation of basic rate pacing from special modes or abnormal pacemaker behavior.

pacing will respond synchronously to the sensed atrial activity in a 1:1 tracking mode. Therefore, the ventricular pacing rate during atrial 1:1 tracking may be any value between the lower rate limit (LRL) and the upper rate limit (URL).

3. *Upper rate behavior*

When the spontaneous atrial rate (SAR) exceeds the upper rate limit (URL), the behavior of the pacemaker will depend upon the settings of the upper rate limit (URL) and the total atrial refractory interval (TARI). Indeed, the shortest spontaneous atrial interval (SAI) that can be sensed is determined by the total atrial refractory interval (TARI), which always consists of two consecutive components, i.e. the AV interval (AVI) and the atrial refractory interval (ARI). The relationship between the total atrial refractory interval (TARI) and the upper rate interval (URI; URI = 60 000: URL) will determine whether the upper rate responses will be a block response or Wenckebach behavior [2].

(a) *Block response.* When the total atrial refractory interval (TARI) equals the upper rate interval (URI), block response will occur whenever the spontaneous atrial rate (SAR) exceeds the upper rate limit (Figure 21). The atrial rate at which block occurs is the blocking onset rate (BOR; BOR = 60 000: TARI).

(b) *Wenckebach behavior.* At atrial rates exceeding the upper rate limit (URL) and lower than the blocking onset rate (BOR), a Wenckebach behavior will occur in a DDD pacemaker programmed with a total atrial refractory interval (TARI) shorter than the upper rate interval (URI).

A pacemaker Wenckebach cycle is characterized by a progressive prolongation of the interval between a sensed P wave and the successive ventricular stimulus until a P wave falling within the atrial refractory interval (ARI) is not sensed and, hence, not followed by a ventricular stimulus resulting in a pause (Figure 22). The P ventricular stimulus interval is the sum of the programmed AV interval (AVIp) and the atrioventricular extension period (AVE) [3]. The atrioventricular extension period (AVE) may range from 0 to the Wenckebach interval (WI) defined as the difference between the upper rate interval (URI) and the total atrial refractory interval (TARI):

$$\mathrm{WI} = \mathrm{URI} - \mathrm{TARI}.$$

Unlike the naturally occurring Wenckebach phenomenon, the interval between consecutive paced ventricular complexes before the pause in the artificial pacemaker Wenckebach, remains constant and equals the upper rate interval (URI). At atrial rates beyond the blocking onset rate (BOR), the pacemaker will again respond with a block response. An overview of the ventricular pacing rates during DDD pacing is depicted in Figure 23.

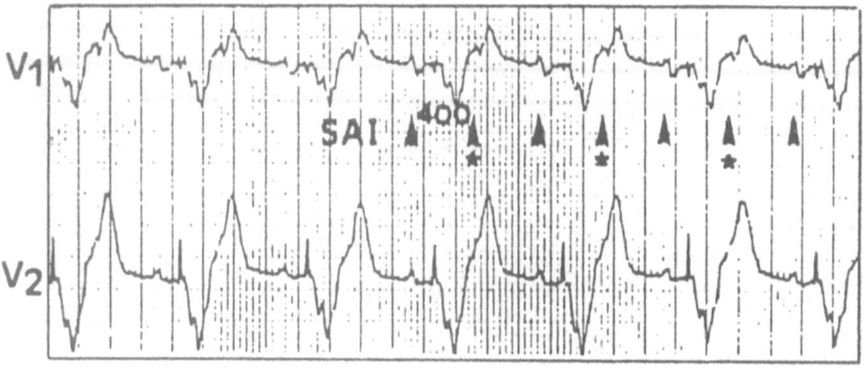

Figure 21. Upper rate behavior in a DDD pacemaker programmed TARI = URI; Block response. DDD pacemaker programmed at a lower rate limit (LRL): 70 ppm; upper rate limit (URL): 120 ppm; upper rate interval (URI = 60 000/120 = 500 ms); AV interval: 175 ms; atrial refractory interval: 325 ms; TARI = URI = 500 ms.
The spontaneous atrial rate (SAR) is 150 beats per minute, i.e. spontaneous atrial interval (SAI = 60 000/150 = 400 ms).The spontaneous atrial rate exceeds the upper rate limit (URL) and pacemaker 2:1 block occurs. Only one (arrow) out of two P waves is sensed which triggers a ventricular output 175 ms after sensing of the P wave. The P waves, indicated by an arrow and asterisk, fall within the atrial refractory interval and are not sensed.

4. *Endless loop tachycardia*

With the advent of DDD pacemakers, a new form of pacemaker mediated tachycardia has emerged, namely the 'endless loop tachycardia'. As in the naturally occurring reentry tachycardias, this artificial circus movement tachycardia requires the presence of two pathways. The anterograde limb is composed of the pulse generator, while the retrograde limb is provided by the potential retrograde conduction pathways in the heart (Table 16).

The endless loop tachycardia (ELT) starts with the sensing of a retrograde P wave which then triggers the next ventricular paced beat. This mechanism perpetuates itself and the tachycardia may persist for several beats or may become sustained [4].

The risk for an endless loop tachycardia exists whenever there is retrograde atrial activation with a ventriculoatrial interval (VAI) which is longer than the postventricular atrial refractory interval (ARI). The shortest endless loop interval (SELI) is determined by the sum of the AV interval (AVI) and the ventriculoatrial interval (VAI):

$$\text{SELI} = \text{VAI} + \text{AVI}.$$

The ventricular pacing rate of the endless loop tachycardia is either determined by the upper rate interval (URI) or by the shortest endless loop

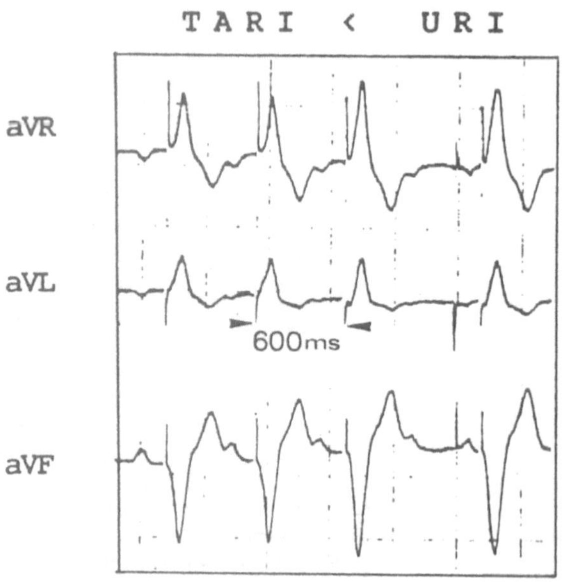

Figure 22. Upper rate behavior of DDD pacemaker programmed TARI < URI: Wenckebach behavior. DDD programmed at a lower rate limit (LRL) 70 ppm; lower rate interval (LRI): 860 ms; upper rate limit 100 ppm; upper rate interval (URI): 600 ms; AV interval: 150 ms; atrial refractory interval: 325 ms; TARI (475 ms) ⟨ URI (600 ms).
The spontaneous atrial rhythm (SAR) is 105 bpm and is higher than the upper rate limit (URL) 100 ppm. The first three P waves are sensed and followed by ventricular stimulation. There is a progressive prolongation of the P to ventricular stimulus interval. The inter-ventricular stimulus interval of the first three paced QRS complexes is 600 ms and equals the upper rate interval (URI). The fourth P wave falls within the atrial refractory interval (ARI) and is not sensed, resulting in a pause in the 5:4 Wenckebach cycle.

Table 16. Endless loop tachycardia (ELT).

PM analog of a natural reentry tachycardia:	
– *Antegrade* artificial pathway	
– pacemaker	AVI
– *Retrograde* natural pathway	
– retrograde VA conduction	VAI

interval (SELI). The largest value of the two will always be the one governing the rate of the endless loop tachycardia (Table 17). The tachycardia cycle length of the endless loop tachycardia (ELT) will equal the shortest endless loop interval (SELI) whenever the sum of the AV interval (AVI) and the

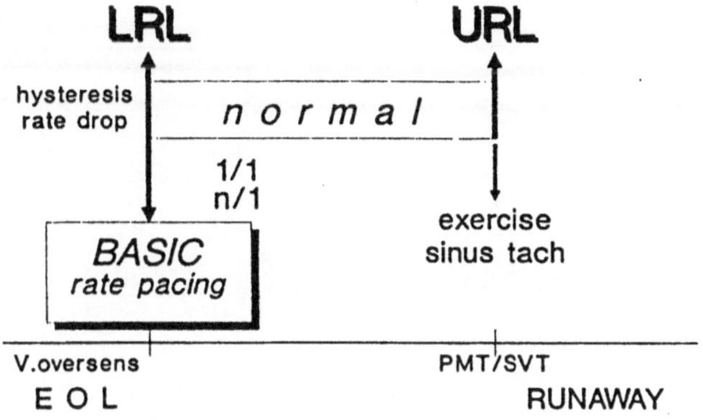

Figure 23. Diagram showing the different ventricular pacing rates during pacemaker operation in the DDD mode.
LRL: lower rate limit; URL: upper rate limit; EOL: end of life criteria; PMT: pacemaker mediated tachycardia; SVT: supraventricular tachycardia; 1:1 = atrial tracking; n/1: pacemaker block response.

ventriculoatrial interval (VAI) is longer than the upper rate interval (URI). An endless loop tachycardia at a rate below the upper rate limit has also been called a balanced endless loop tachycardia (BELT). When SELI (the sum of VAI and AVI) is shorter than the upper rate interval (URI), this latter will limit the rate of the endless loop tachycardia to the upper rate limit (URL) of the pacemaker. As the ventricular stimulus will not be released until the upper rate interval (URI) has been timed out, the programmed AV interval will be prolonged by a waiting period called the atrioventricular extension period (AVE). Consequently, the ventriculoatrial conduction time can be calculated [5] by subtracting the atrioventricular interval (AVI) from the tachycardia cycle length only when the rate of the endless loop tachycardia occurs below the upper rate limit (URL).

The common denominator for the *initiation* of an endless loop tachycardia (Table 18) is the presence of a long AV interval (AVI) and a short postventricular atrial refractory interval (ARI). An endless loop tachycardia may begin following retrograde conduction of premature ventricular complex (Figure 24) or by an atrial premature complex and a P wave that occurs retrogradely after a normal antegrade P wave. The endless loop tachycardia may also start by a ventricular paced beat after a noncaptured atrial stimulus or after loss of atrial sensing, electromagnetic or electromyographic interference, or by an asynchronous ventricular stimulus elicited by chest-wall stimulation or by application of a magnet over the pacemaker [5].

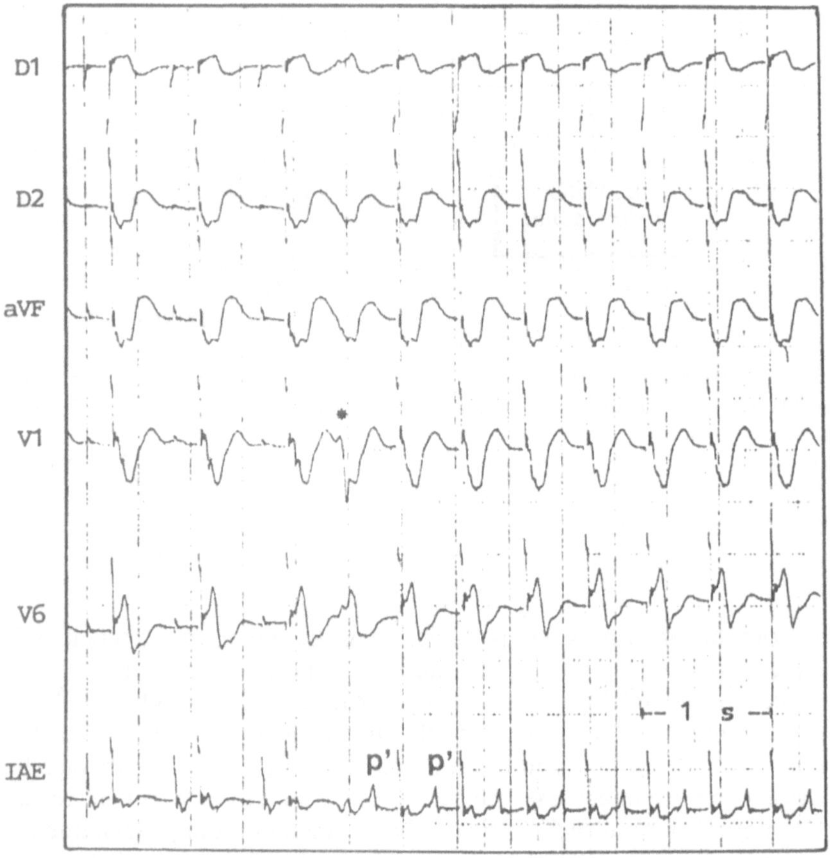

Figure 24a.

Figure 24. Endless loop tachycardia initiated by a premature ventricular complex. DDD pacemaker temporarily programmed in the electrophysiologic laboratory at a lower rate limit (LRL): 88 ppm; AV interval (AVI): 180 ms; upper rate limit (URL): 140 ppm; postventricular atrial refractory interval (ARI): 250 ms.

(a) ECG leads D1, D2, AVF, V1, V6 and intra-atrial electrogram (IAE); paper speed 25 mm/s. The fourth beat is a premature complex (asterisk) which initiates a balanced endless loop tachycardia at a rate of 130 ppm. The premature ventricular complex is followed by retrograde atrial activation (P') 280 ms after the ventricular stimulus. Since the retrograde P wave falls outside the atrial refractory interval (ARI) of 250 ms, the P wave is sensed and

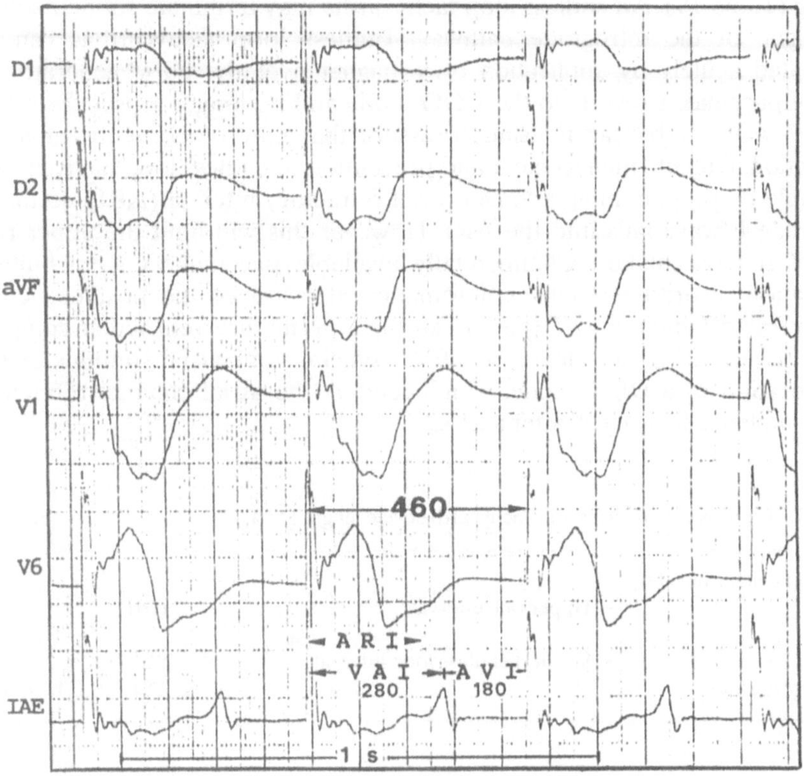

Figure 24b.

followed by a ventricular output after an AV interval (AVI) of 180 ms. The ventricular paced beat is, in its turn, followed by a retrograde P wave and the endless loop tachycardia becomes sustained.

(b) ECG leads D1, D2, AVF, V1, V6 and intra-atrial electrogram (IAE); paper speed 100 mm/s; recorded during a sustained endless loop tachycardia at a rate of 130 ppm. The tachycardia cycle length is 460 ms. The shortest endless interval (SELI) equals the sum of ventriculoatrial interval (VAI: 280 ms) and the AV interval (AVI: 180 ms) and is longer than the upper rate interval (URI: 430 ms).

Mechanisms that *terminate* an endless loop tachycardia have in common that they result either in a loss of ventriculoatrial conduction or a loss of atrial sensing (Table 19). An endless loop tachycardia may terminate spontaneously by fatigue of the retrograde pathway. Endless loop tachycardias can be terminated acutely by application of a magnet over the pulse generator so that the pacemaker reverts to the DOO mode and no longer senses the retrograde P wave or by carotid sinus pressure that may block ventriculoatrial conduction. Definite management and prevention of endless loop tachycardias is feasible by programming a longer atrial refractory interval (ARI), so that a retrograde P wave falls into the ARI. However this will limit the upper rate limit of the pacemaker. Commercially available pacemakers have built-in termination algorithms such as automatic extension of the postventricular atrial refractory interval (ARI) after sensing a premature ventricular complex (PVC) (Figure 25), the release of a PVC synchronous atrial stimulus or the incorporation of a DDX mode, which renders the atrial channel refractory for a complete pacing cycle after a PVC.

Table 17. Endless loop tachycardia: tachycardia cycle length (TCL).

TCL will be determined by:		
	— Upper rate interval	URI
or		
	— Shortest endless loop interval	SELI
whichever is greater		$SELI = VAI + AVI$

Table 18. Endless loop tachycardia: initiating mechanisms (short ARI; long AVI).

— PVC with retrograde conduction	— Myopotentials/Electromagnetic interference
— Premature atrial contraction	— Application of magnet
— Loss of atrial capture	— Chest wall stimulation
— Undersensing of P waves	— WB pacemaker behavior/AV extension

Table 19. Endless loop tachycardia: termination (loss of VA conduction; loss of atrial sensing).

- Application of magnet: by reversion to DOO mode
- Carotid sinus pressure: by blocking VA conduction
- Reprogramming longer ARI
- Automatic termination algorithm:
  - post PVC extension of ARI
  - PVC synchronous atrial stimulus
  - DDX mode

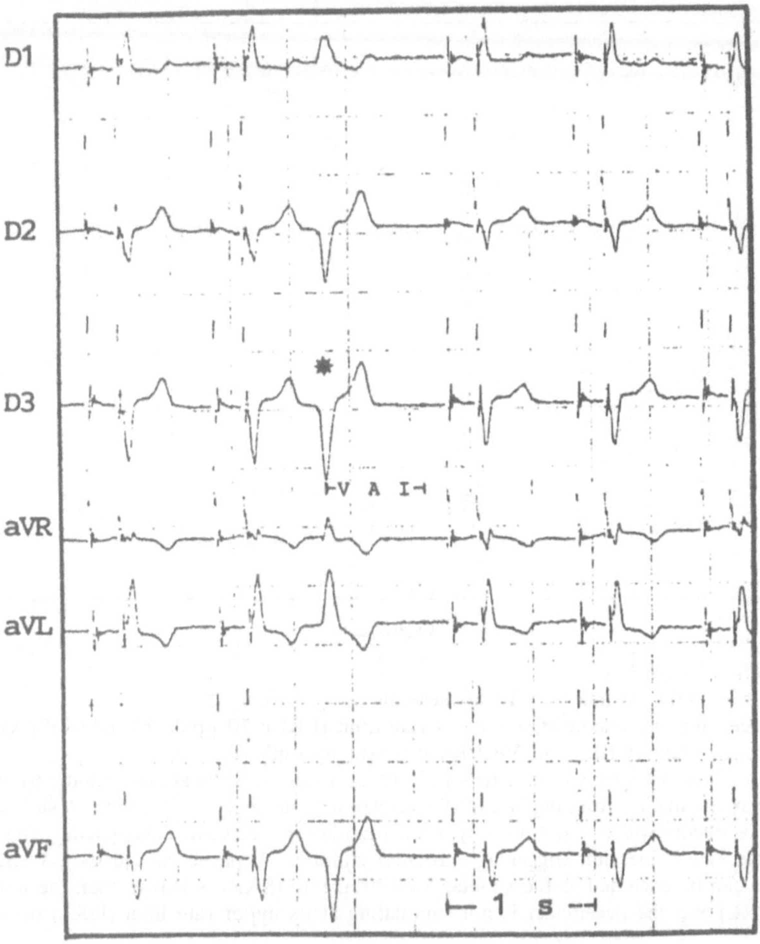

Figure 25. Automatic extension of the postventricular atrial refractory period after sensing of a premature ventricular complex. DDD pacemaker programmed at a lower rate (LRL): 70 ppm; AV interval (AVI): 200 ms.
The first two complexes demonstrate DDD lower rate operation at a rate of 70 ppm. The third complex (asterisk) is a premature ventricular complex (PVC) that might give rise to retrograde activation of the atrium and start an endless loop tachycardia. A PVC is recognized by the pacemaker as a sensed ventricular event not preceded by a sensed or paced atrial event. Some commercially available pacemakers have a built-in algorithm to prevent the initiation of an endless loop tachycardia. This particular pacemaker prolongs the postventricular atrial refractory interval (ARI) after sensing a PVC. This mechanism results in a pause after the PVC that is longer than the normally expected ventriculoatrial interval (VAI).

CASE STUDIES

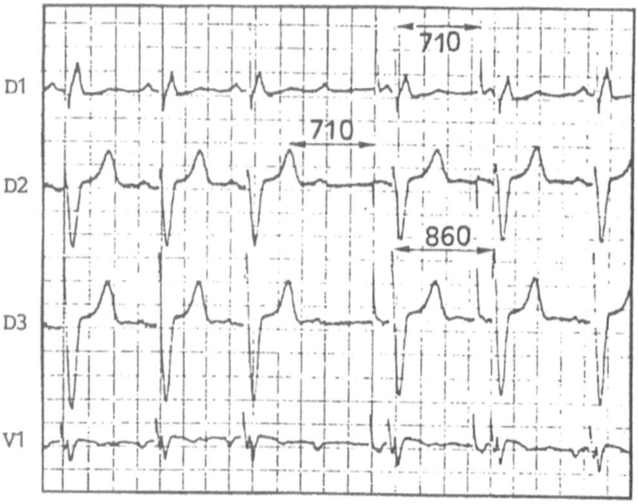

Figure 26.

Case study 1: Figure 26.
Inhibition of DDD pacemaker by T wave sensing.
DDD pacemaker programmed at a lower rate limit (LRL): 70 ppm; AV interval (AVI): 150 ms: Upper rate limit (URL): 130; Ventricular sensitivity 1 mV.

There is a spontaneous atrial rhythm (SAR) at a rate of 80 beats per minute followed by ventricular stimulation resulting in a QRS complex with normal configuration. A sudden pause of 860 ms occurs which is followed by atrial stimulation and ventricular pacing. The P wave within the pause does not trigger a ventricular stimulus. A pause caused by a Wenckebach behavior can be excluded as the spontaneous atrial rate (SAR) is slower than the upper rate limit (URL) and the pacemaker is not stimulating at its upper rate limit (URL) prior to the pause.

After the pause, the DDD pacemaker is stimulating the atrium i.e. lower (basic) rate pacing. The calculated ventriculoatrial interval (VAI) equals the lower rate interval (LRI) minus the programmed AV interval (AVI); VAI = LRI − AVIp; 710 ms = 860−150.

The ventriculoatrial interval (VAI) can be measured by putting one leg of the calipers on the ventricular stimulus following the pause and the other leg on the successive atrial spike. The pair of calipers is then set with the right leg on the first atrial spike following the pause and by bringing over backwards the VAI, the left leg of the calipers falls on the T wave. This means that the ventricular channel has sensed the T wave which has reset the pacemaker. Resetting of the ventricular channel of the pacemaker starts a new postventricular atrial refractory period (ARI) and, therefore, renders the pacemaker unable to sense the P wave. Malsensing of the pacemaker by undersensing of the P wave can be excluded as this will not result in a pause.

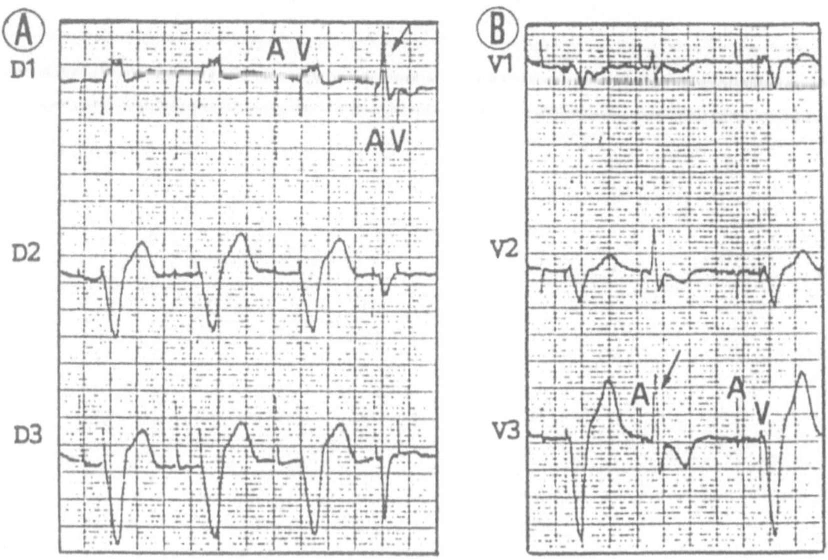

Figure 27.

Case study 2: Figure 27
Atrial fibrillation:
DDD pacemaker programmed at a lower rate limit (LRL): 80 ppm; AV interval (AVI): 180 ms; Ventricular blanking period: 50 ms.

(a) *Right panel:* ECG lead D1, D2, and D3. There is noncapture of the atrial stimulus (A) because of atrial fibrillation. Normal ventricular pacing is present during the first three complexes. The fourth QRS complex (arrow) is a spontaneous QRS complex and occurs 30 ms after the atrial stimulus. As the QRS complex falls within the ventricular blanking period of 50 ms, it is not sensed by the ventricular channel. The ventricular stimulus does not capture the ventricle, since it falls within the ventricular refractory period.

(b) *Left panel:* ECG lead V1, V2 and V3. Due to atrial fibrillation, there is noncapture of the atrium. The second QRS complex (arrow) is a spontaneous QRS complex and occurs 120 ms after the atrial stimulus. As it occurs outside the ventricular blanking period, the QRS complex is sensed and the ventricular output is inhibited.

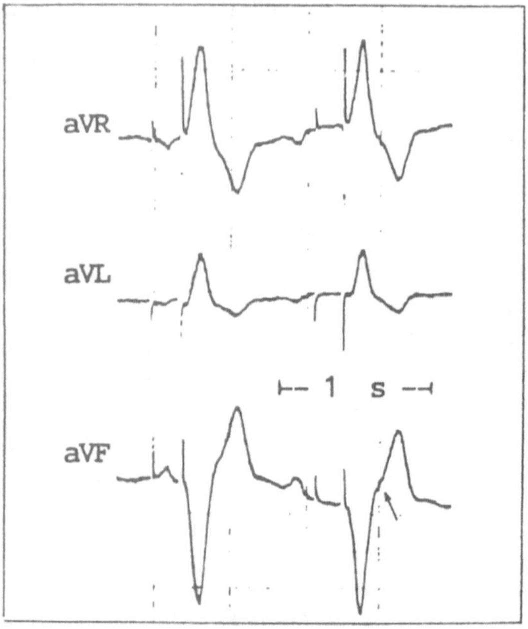

Figure 28.

Case study 3: Figure 28
Atrial undersensing.
DDD pacemaker programmed at a lower rate limit (LRL): 70 ppm; AV interval (AVI): 150 ms.

The first stimulus captures the atrium and is followed after an AV interval of 150 ms by a ventricular stimulus which captures the ventricle. The next P wave is not sensed and is followed by an ineffective atrial stimulus, since it falls within the atrial refractory period. After the programmed AV interval, a ventricular stimulus captures the ventricle. The ST segment of the second QRS complex is distorted by a retrograde P wave (arrow). This could start an endless loop tachycardia.

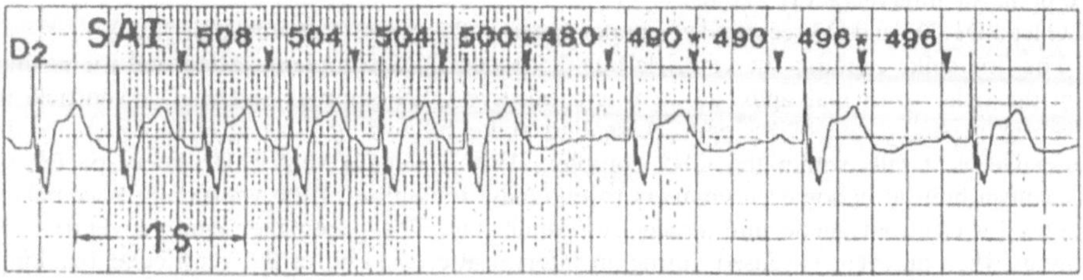

Figure 29.

Case study 4: Figure 29

Upper rate behavior of DDD pacemaker: Transition from 1:1 atrial tracking to 2:1 block response during exercise.

DDD pacemaker programmed at a lower rate limit (LRL): 60 ppm; Lower rate interval (LRI): 1000 ms; Upper rate limit (URL): 120 ppm; Upper rate interval (URI): 500 ms; AV interval (AVI): 175 ms; Atrial refractory interval (ARI): 325 ms; TARI = URI = 500 ms.

The spontaneous atrial rate (SAR) increases progressively from 118 bpm to 125 bpm, i.e. spontaneous atrial interval (SAI) shortens from 508 to 480 ms. There is atrial 1:1 tracking as long as the spontaneous atrial interval (SAR) is lower than the upper rate limit (URL) of 120 ppm. As the atrial rate (SAR) exceeds the upper rate limit (URL), i.e. spontaneous atrial interval (SAI) shorter than the upper rate interval (URI) of 500 ms, 2:1 block occurs and the ventricular pacing rate drops from 120 to 60 ppm.

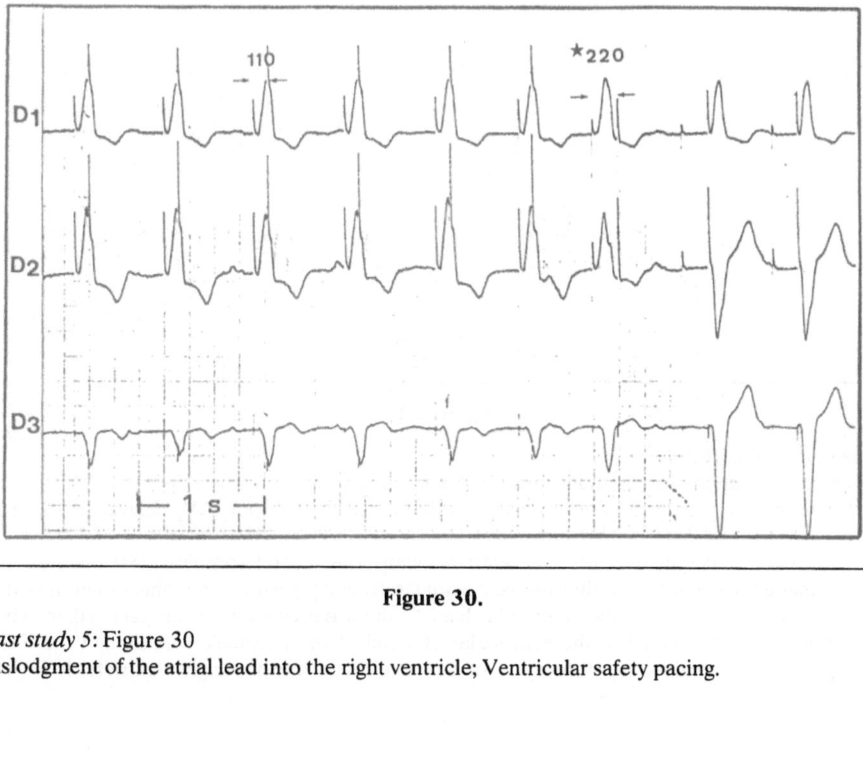

Figure 30.

Cast study 5: Figure 30

Dislodgment of the atrial lead into the right ventricle; Ventricular safety pacing.

DDD pacemaker programmed at a lower rate limit (LRL): 70 ppm; Upper rate limit (URL): 150 ppm; AV interval (AVI): 220 ms.
ECG lead D1, D2 and D3 recorded at a paper speed of 25 mm/s. (Retouched original).

The first spike captures the ventricle but the paced QRS complex is abnormal for pacing from the right ventricular apex, as there is a positive QRS complex in D2. The ventricle is driven at a pacing rate of 85 ppm. A second spike follows 110 ms after the first and is ineffective as it falls within the QRS complex. This is due to ventricular safety pacing, a mechanism built-in in some pacemakers to eliminate inhibition of the ventricular pacemaker channel by electromagnetic interference, crosstalk, etc. The pacemaker emits a ventricular stimulus when an event is sensed during an interval after the end of the ventricular blanking period.

There are P waves at a regular interval which are not sensed.

The programmerhead is placed over the pacemaker which converts the pacemaker from the DDD to the DOO mode and the atrial output of the pacemaker is diminished until atrial subthreshold pacing occurs. The first QRS complex (asterisk) in the DOO mode is preceded by a pacemaker artifact. This stimulus is smaller than in the previous complexes because of the reduction of the atrial output pulse but still captures the ventricle. A spike is following the QRS complex 220 ms after the first spike at the programmed AV interval. Ventricular safety pacing no longer occurs as the pacemaker is operating in the DOO mode and is not capable of sensing. The last two QRS complexes are preceded by a small atrial pacemaker artifact that does not capture the atrium because of atrial subthreshold pacing at a magnetic rate of 85 ppm. The ventricular stimulus captures the ventricle and the paced QRS complex in D2 is now negative. This proves that the atrial lead has been dislodged into the right ventricle.

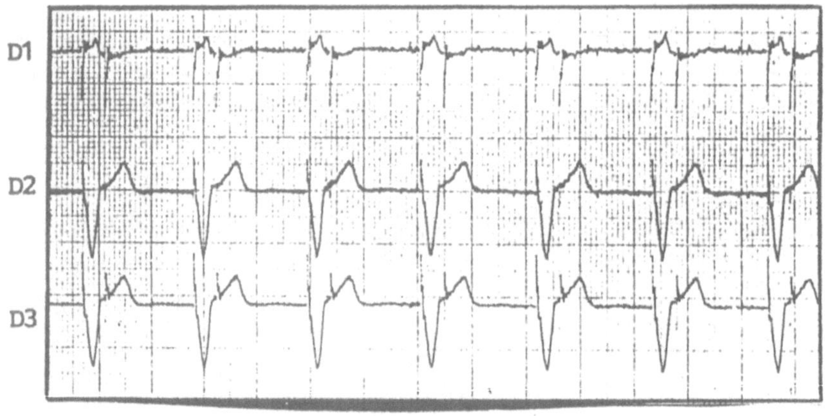

Figure 31.

Case study 6: Figure 31
Reversed connection of atrial and ventricular lead.
DDD pacemaker programmed at a lower rate limit (LRL): 70 ppm; AV interval (AVI): 180 ms.

The first stimulus depolarizes the ventricle, while the second stimulus, 180 ms after the first, is ineffective as it falls within the ventricular refractory period. This phenomenon is due to a reversed connection of the ventricular lead to the atrial channel of the pacemaker, while the atrial lead is connected to the ventricular channel of the pacemaker. Although the ECG tracing could be obtained by dislocation of the atrial lead into the right ventricle, it is very unlikely as the configuration of the paced QRS complex is normal. A chest x-ray and/or temporary programming of the pacemaker into the AOO or VOO mode may be necessary to make a correct diagnosis.

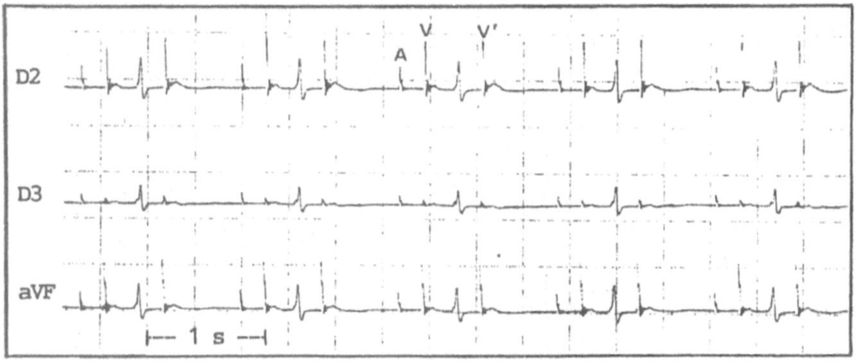

Figure 32.

Case study 7: Figure 32

Presence of three pacing stimuli during DDD pacing.

DDD pacemaker programmed at a lower rate limit (LRL): 70 ppm; AV interval (AVI): 200 ms; Atrial & ventricular output 2 V, 0.2 ms.

Three stimuli are present, two preceding and one following the QRS complex. The leads have been reversely connected during implantation. The first spike is the output stimulus of the atrial channel (A) which is connected to the ventricular lead. The first spike does not capture the ventricle due to subthreshold pacing. The second spike is the output of the ventricular channel (V) of the pacemaker which is connected to the atrial lead. The second spike following 200 ms after the first spike captures the atrium which conducts to the ventricle, resulting in a narrow QRS complex. The spontaneous QRS complex is sensed by the ventricular lead which is connected to the atrial channel of the pacemaker. After an AV interval (AVI) of 200 ms, the ventricular channel emits a stimulus (V') which is transmitted via the lead located in the atrium. It is not evident from the ECG tracing, whether the third stimulus results in capture of the atrium.

REFERENCES

1. Mond H (1983) *The Cardiac Pacemaker: Function and Malfunction.* New York: Grune & Stratton
2. Stroobandt R, Willems R, Holvoet G, Backers J, Sinnaeve A (1986) Prediction of Wenckebach behavior and block response in DDD pacemakers. PACE 9: 1040—1046
3. Willems R, Stroobandt R, Holvoet G, Backers J, Sinnaeve A (1987) Calculation of AV extension during pacemaker Wenckebach. PACE 10: 763
4. Barold SS (1985) *Modern Cardiac Pacing,* Mount Kisco, New York: Futura Publishing
5. Furman S, Hayes D, Holmes D (1989) *A Practice of Cardiac Pacing,* 2nd revised and enlarged edn. Mount Kisco, New York: Futura Publishing

6. Rate-adaptive cardiac pacing

A. JOHN CAMM & CLIFFORD J. GARRATT

Any artificial pacing device that is capable of increasing its pacing rate with exercise or other physiological stimuli, can be considered a rate-adaptive or rate-responsive device. The importance of increased heart rate for augmenting cardiac output with exercise has been clearly documented [1], and the benefits of rate-adaptive systems over fixed rate pacemakers have been documented repeatedly in terms of exercise tolerance, symptomatology, and haemodynamic variables [2, 3].

In order to achieve rate response, pacemakers have been designed to track atrial activity (atrial dependent) or to adjust their rate of discharge according to a more remote, but sometimes more accurate indicator of the most 'appropriate' sinus rate (atrial independent).

ATRIAL-INDEPENDENT ADAPTIVE-RATE PACEMAKERS

Sensors of metabolic demand that are independent of atrial rate can be divided into 5 main groups.

1. *Vibration sensing*

The sensors incorporated in the most commercially successful rate-adaptive pacemakers are those that detect bodily motion or activity. Body motion can be detected as vibrations by a piezoelectric crystal bonded to the inside of the pulse generator casing (Activitrax, Medtronic). Other vibration-sensing pacemakers include Sensolog (Siemens) and Dash (Intermedics). Vibration sensing pacemakers provide a fast response time, are robust and no special lead is required. In a number of activities, however, it is clear that bodily vibration is not well matched to metabolic activity and pacing-rate response to increasing gradients, swimming, or cycling may be inappropriate. At the same time, inappropriate symptomatic pacing tachycardias can result from train (Figure 1) and air travel, particularly in helicopters [4]. Clearly, this

E. Andries, P. Brugada & R. Stroobandt (eds.), How to face 'the faces' of cardiac pacing, 119–126.

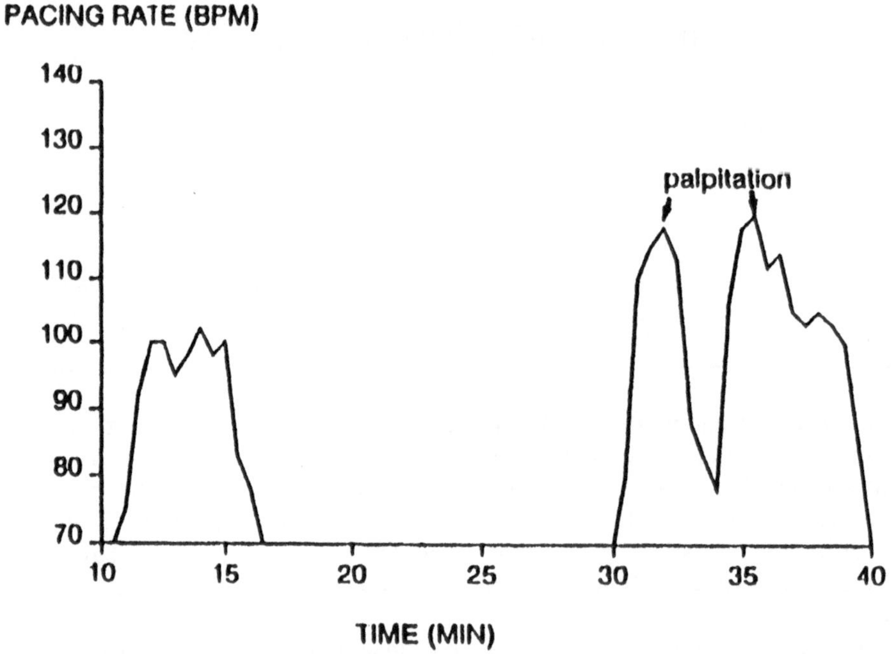

Figure 1. Effect of train journey on Activitrax pacemaker.

form of pacing system is inappropriate in patients likely to be exposed to excessive bodily vibration (vibrating tools etc).

2. *Impedance sensing*

A second major group of rate-adaptive atrial independent pacemakers are those that use impedance measurements as a sensor. Transthoracic electrical impedance rises with expiration and falls with inspiration, the amplitude of these changes being related to tidal volume. The minute ventilation sensing pacemaker (Meta MV, Telectronics) uses a single bipolar electrode in the right ventricle (or atrium) and impedance is measured by injecting current between the ring electrode and the pacemaker casing [5] (Figure 2). Rate response is well correlated with the degree of exertion during both graded and burst exercise. Potential problems with these devices relate to interference by talking during exercise (attenuated response) or arm swinging (increased rate response). The influence of cardiorespiratory disease on the rate response of these pacemakers has not been formally assessed but is likely to be small. Minute ventilation is not the only sensor of metabolic activity that can be derived from impedance measurements. Stroke volume,

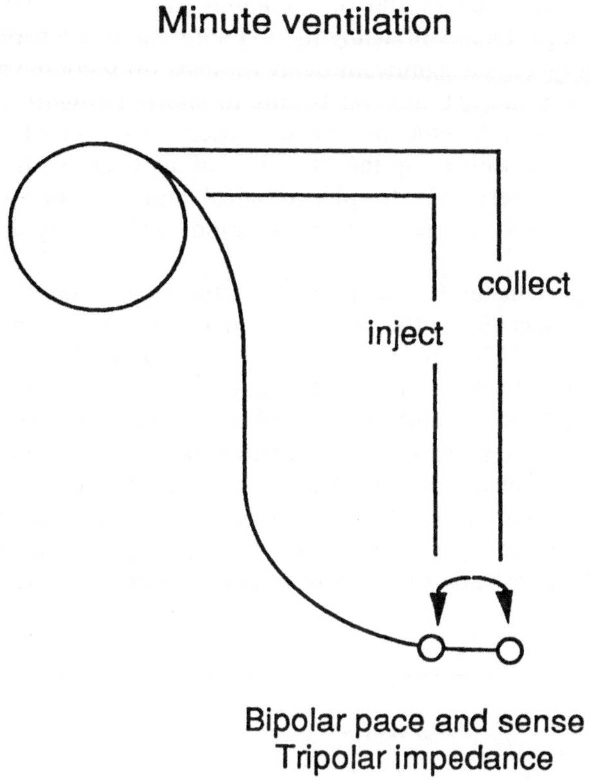

Figure 2. Meta MV pacemaker.

rate of change of stroke volume, and preejection interval can be derived from changes in impedance measured in the right ventricle. The estimation of changes in these variables forms the basis for a rate-adaptive pacemaker (Precept, CPI). At present, clinical experience with this device is very limited.

3. *Sensing of the paced evoked response*
The development of a method of reducing electrode polarisation has allowed paced evoked repolarisation and depolarisation potentials to be detected by a conventional unipolar electrode system [6]. Using this method, the duration of the local endocardial equivalent of the QT interval, known as the stimulus-T interval, can be calculated. Rate adaptive pacing systems have been developed [7] that exploit the known effects of exercise upon the QT interval to calculate an appropriate pacing rate response (Rhythmyx, Vitatron). It is now recognised that exercise shortens the QT interval (and the stimulus-T inter-

val) independently of changes in heart rate, probably as a result of changes in the levels of sympathetic tone and/or circulating cateholamines [8, 9].

The principal problem with using the QT interval as a sensor of metabolic demand is that there is a significant delay (at least 60 seconds) after initiation of exercise before the QT interval begins to shorten (Figure 3) [10], and a similar delay before lengthening occurs after cessation of exercise. The principal advantage with using the QT interval as a sensor, however, is that this measurement responds to physiological stimuli other than dynamic exercise and there is an excellent correlation with steady-state metabolic demand [11].

The Prism pacemaker (Cordis) also uses the ventricular-evoked depolarisation, or more specifically the ventricular depolarisation gradient (VDG) to control pacing rate [12]. The VDG is derived from an electronic integration of the paced evoked QRS complex: this function remains constant at rest but an increase in metabolic demand will lead to a reduction in the VDG. Heart rate increase, however, lead to an increase in the VDG such that, under physiological circumstances, the changes are exactly matched, maintaining the VDG at a constant level. The Prism system uses an algorithm that attempts to maintain a constant VDG by varying the pacing rate: early clinical results suggest that this system provides appropriate rate response to

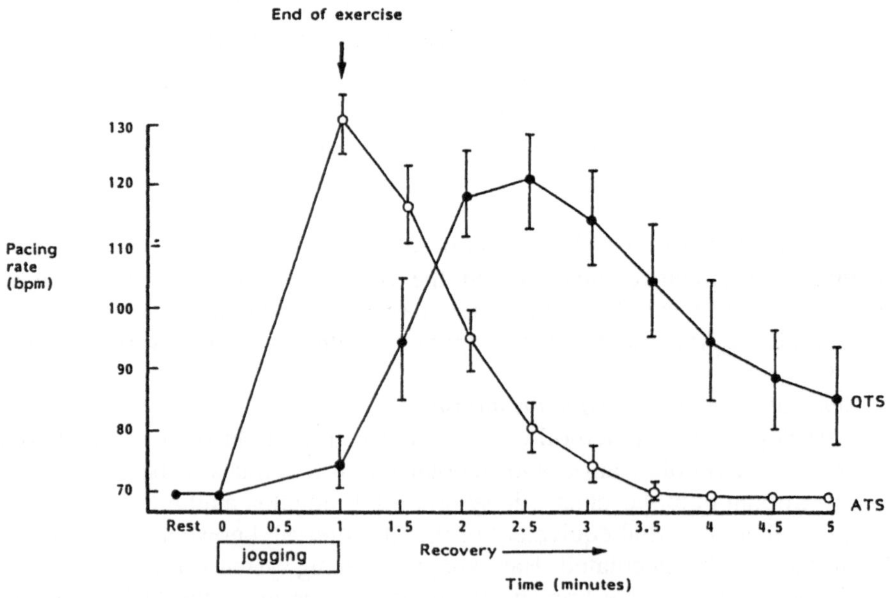

Figure 3. Delay in pacing responses of QT pacemaker to exercise.

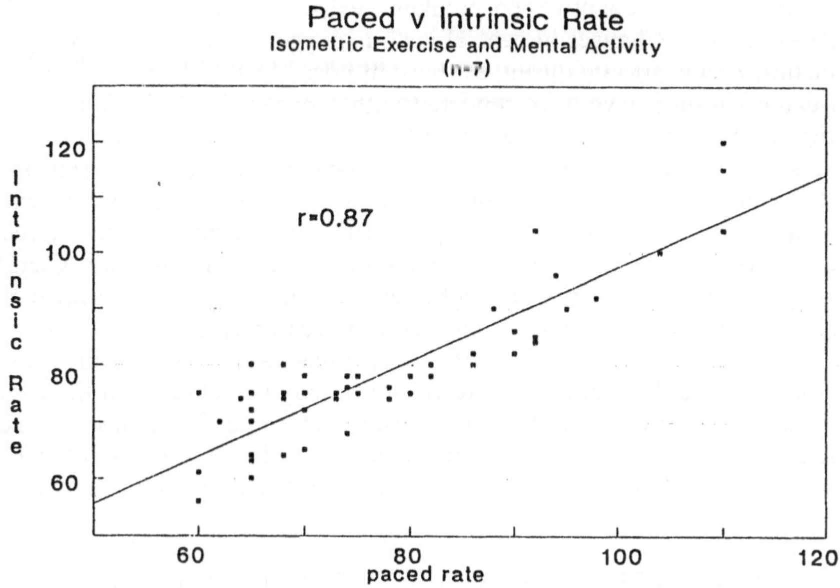

Figure 4. Comparison of paced and intrinsic sinus rate in patients with the Prism pacemaker.

exercise and mental activity (Figure 4). However, pacing rate response is not always appropriate to other physiological stimuli. In particular, changes which increase ventricular volume, such as adopting a supine posture, also reduce the VDG and causes an inappropriate heart rate increase.

4. *Temperature sensing*
Right ventricular temperature is known to rise with exercise [13] and this increase in temperature has been shown to correlate well with the normal sinus response, particularly at high levels of exercise. The fact that right ventricular temperature falls at the very start of exercise (due to return of cool venous blood from the peripheries) has been used to trigger the start of rate response of temperature-sensing rate-adaptive pacemakers. Early clinical results suggest that a response time of 20—30 seconds is achievable. Two temperature-sensing devices are currently available or undergoing clinical evaluation: Thermos (Biotronik) and Kelvin 500 (Cook). A number of potential problems exist with the use of temperature as a sensor of metabolic activity: it is uncertain whether the temperature response will be appropriate during submaximal exercise, and the initiation of exercise in a patient who is recovering from previous exercise does not cause an initial dip in temperature. Further clinical evaluation of these devices is awaited.

5. *Miscellaneous sensors*

A number of rate adaptive pacemakers exist utilising other sensors of metabolic activity. Changes in central venous oxygen saturation show a good correlation with heart rate during exercise in normal subjects, and good short term clinical results have been reported with a pacemaker using an oxygen-sensing device as the metabolic sensor [14].

Pressure changes in the right ventricle have been used as a sensor of metabolic demand, using a piezoelectric crystal lying within the right ventricular cavity for measurement of d*P*/d*t* (DPDT, Medtronic). With exercise, there is rapid acceleration of the pacing rate in the first minute. Clearly, pathological changes affecting the right ventricle may affect the performance of this pacemaker and long-term clinical results are awaited.

It is clear from the above that each sensor has its own flaws and benefits and it is probable that improved rate-response may be achieved in the future by the combination of more than one sensor. Future combinations are likely to include a sensor with rapid response (such as a vibration-sensor) together with a more 'metabolic' sensor such as evoked response or minute ventilation.

ATRIAL-DEPENDENT ADAPTIVE RATE PACEMAKERS

Ventricular pacing rate can be dependent on atrial rate directly (i.e. with maintenance of AV synchrony) or indirectly, using measurement of atrial activity over a given time period. Several reports have documented the haemodynamic advantages offered by atrial synchronous as compared with fixed-rate ventricular pacing but improvement of exercise haemodynamics and performance is related primarily to increased rate rather than the presence of AV synchrony. The question arises, therefore, as to whether there are any advantages of atrial-dependent over atrial-independent rate adaptation.

Preliminary comparisons of atrial dependent (VDD) and atrial independent (VVIR) rate-adaptive pacing in terms of haemodynamic changes show a small trend in favour of the atrial-dependent modes but, in general, such differences are unlikely to be noticed by the patients themselves. A notable exception to this is that group of patients suffering from the pacemaker syndrome during VVIR mode. In these patients, VVIR pacing is clearly contraindicated. Perhaps more important than the small haemodynamic advantages offered of VVIR pacing [15]. All the increased risks of fixed rate VVI pacing, such as induction of atrial fibrillation and incidence of thromboembolism, are present with VVIR pacing. In patients with sick sinus syndrome, constant-rate atrial pacing (with AV synchrony) is associated with a lower incidence of heart failure, thromboembolism and mortality than those with constant-rate ventricular pacing. Loss of AV synchrony and retrograde VA conduction have been implicated in the development of congestive failure in these patients. AV asynchrony results in a greater release of catecholamines and has been shown to cause left ventricular wall-motion abnormalities. For these reasons it is recommended that, in the presence of normal sinus node function and AV block, that atrial-dependent adaptive rate pacing is the pacing mode of choice.

Atrial-independent adaptive rate pacing is clearly the preferable mode in patients with permanent or almost permanent atrial arrhythmias. However, there is evidence to suggest that AV synchronisation may protect against the development of atrial fibrillation in patients with paroxysmal atrial fibrillation and therefore a history of this arrhythmia is not a contraindication to VDD pacing. VVIR pacing is the required mode in patients with giant atria (difficult to fix an atrial lead) or silent atria (inability to pace or sense).

Patients with an inappropriate sinus node response to exercise (chronotropic incompetence) in addition to heart block pose a particular problem in terms of pacemaker selection. It is clear that atrial-dependent rate adaptive pacing is likely to be of little use if the sinus node itself does not adapt to the changes associated with exercise. In this situation the DDDR mode of pacing is to be preferred [16]: in this mode the atrial pacing rate is sensor driven, allowing adaptation of rate to exercise whilst maintaining AV synchrony. An additional advantage of the DDDR pacemakers in current use is the availability of a number of pacing modalities in one unit. As conduction and rhythm changes cannot always be predicted, the choice of a standard pacemaker is tailored to suit the condition of the patient at the time of implant. DDDR pacemakers potentially have the optimal pacing mode for all situations and, as such, once implanted, require at most reprogramming to adapt to any new situation.

CONCLUSIONS

Rate-adaptive pacing of any form is preferable to fixed rate ventricular pacing in terms of symptoms, haemodynamics, and exercise tolerance. If permanent atrial fibrillation is present VVIR pacing is required. If sinus node response to exercise is normal, VDD pacing is the preferred pacing mode: VVIR pacing may be associated with an increase in long-term morality and a DDDR mode is unnecessary. In the presence of chronotropic incompetence (inappropriate sinus response to exercise) DDDR pacing is preferable to VDD.

REFERENCES

1. Samet P, Castillo C, Berstein WH. (1966) Hemodynamic sequelae of atrial, ventricular, and sequential atrioventricular pacing in cardiac patients. *Am Heart J* 72: 725—729.
2. Benditt DG, Mianulli M, Fetterr J, et al. (1987) Single-chamber cardiac pacing with activity-initiated chronotropic response: evaluation by cardiopulmonary exercise testing. *Circulation* 75: 184—191.

3. Lipkin DP, Buller N, Frenneaux M, et al. (1987) Randomised crossover trial of rate-responsive Activitrax and conventional fixed rate ventricular pacing. *Br Heart J* 58: 613–616.
4. Toff WD, Leeks C, Joy M, et al. (1987) The effect of aircraft vibration on the function of an activity-sensing pacemaker (abst.). *Br Heart J* 57: 573.
5. Lau CP, Antoniou A, Ward DE, Camm AJ. (1988) Initial clinical experience with a minute-ventilation sensing rate modulated pacemaker: improvements in exercise capacity and symptomatology. *PACE* 11: 1815–1822.
6. Donaldson RM and Rickards AF. (1983) The ventricular endocardial paced evoked response. *PACE* 6: 253.
7. Rickards AF, Norman J. (1981) Relation between QT interval and heart rate. New design of physiological adaptive cardiac pacemaker. *Br Heart J* 45: 56–61.
8. Fananapazir L, Bennett DH, Faragher EB. (1983) Contribution of heart rate to QT interval shortening during exercise. *Eur Heart J* 4: 265–271.
9. Milne JR, Ward DE, Spurrell RAJ, Camm AJ. (1982) The ventricular paced QT interval-effects of rate and exercise. *PACE* 5: 352–358.
10. Mehta D, Lau CP, Ward DE, Camm AJ. (1988) Comparative evaluation of chronotropic responses of QT-sensing and activity sensing rate-responsive pacemakers. *PACE* 11: 1405–1412.
11. Camm AJ, Garratt CJ, Paul V. (1989) Single-chamber rate adaptive pacing. *J Electrophysiol.* 3: 181–189.
12. Paul V, Garratt C, Ward DE, Camm AJ. (1989) Closed loop control of rate adaptive pacing: clinical assessment of a system analysing the ventricular depolarisation gradient. *PACE* 12: 1896–1902.
13. Bazett HC. (1951) Theory of reflex controls to explain regulation of temperature at rest and during exercise. *J Appl Physiol* 4: 245.
14. Stangl K, Wirtsfeld A, Heinz R, Laule M. (1988) First clinical experience with an oxygen saturation controlled pacemaker in man. *PACE* 11: 1882–1887.
15. Camm AJ, Katritsis D. (1990) Ventricular pacing for sick sinus syndrome — a risky business? *PACE* 13: 695–699.
16. Paul V, Garratt CJ, Camm AJ. (1989) Combination of sensors to provide optimal pacing rate response. Clinical Cardiology 7: 400–404.

7. Dual chamber rate responsive (DDDR) pacing: ventricular versus atrial timing

ROLAND STROOBANDT, ROGER WILLEMS &
ALFONS SINNAEVE

Physiologic pacing systems should not only maintain AV synchrony but also respond to the patient's metabolic needs by varying heart rate and, consequently, cardiac output. The addition of rate responsiveness to DDD pacemakers may almost completely mimic normal heart function in patients in whom the sinus node is not a suitable indicator of physiologic heart-rate modulation. As DDD pacemakers come in widespread clinical use, new electrocardiographic appearances will undoubtedly emerge. Correct interpretation of an ECG depends upon adequate familiarity with the device and complete understanding of its principles of operation and its peculiarities. The distinction between normal or abnormal functioning may become clouded in highly sophisticated devices and an unexpected ECG should not always be taken as indicating device failure.

Contemporary DDDR pacemakers may have either atrial or ventricular timing concepts. This also results in a different behavior at the lower rate during tracking and at the upper rate. A complete understanding of the operational concepts may be necessary to correctly interpret DDDR pacemaker electrograms.

LOWER RATE PACING

(a) *Ventricular timing concept*

Most commercially available DDD pacemakers up to now operate on a ventricular timing concept. This means that there is a limit below which the ventricular paced rate is not allowed to drop. The lowest ventricular rate is determined by the ventricular lower rate interval (VLRI), which is the longest interval between two consecutive paced ventricular events or a sensed spontaneous ventricular complex and the subsequent paced ventricular event.

When defining the interventricular interval (IVI) as being the interval between two consecutive ventricular events, then it follows from the ventricular timing concept that the interventricular interval (IVI) can be shorter

E. Andries, P. Brugada & R. Stroobandt (eds.), How to face 'the faces' of cardiac pacing, 127—137.

or equal to the ventricular lower rate interval (VLRI), but never longer (Figure 1).

The interventricular interval consists of two consecutive portions, the atrioventricular interval (AVI) and the pacemaker ventriculoatrial interval (VAI), also called the atrial escape interval (AEI). The atrioventricular interval (AVI) may be shorter than the programmed one when a spontaneous ventricular event interrupts the programmed AV delay.

During lower rate pacing, the pacemaker ventriculoatrial interval (VAI) starts with the ventricular stimulus or ventricular sensed event and ends at the subsequent atrial stimulus. The pacemaker ventriculoatrial interval (VAI) is always constant and equals the ventricular lower rate interval (VLRI) minus the programmed AVI (AVIp) (Table 1). Consequently measurement of the pacemaker ventriculoatrial interval (VAI) is of the utmost importance to verify the normal functioning of the DDD pacemaker at its lower rate.

Exceptionally, when special modes are programmed on, such as, e.g., 'hysteresis' or 'rate drop at night', the pacemaker VAI may become longer than the lower rate interval minus the programmed atrioventricular interval (AVIp). In all other circumstances, a prolongation of the pacemaker ventriculoatrial interval (VAI) during lower rate pacing is abnormal and may indicate either false inhibition or impending battery depletion with slowing of the pacing rate below the ventricular lower rate limit (VLRL).

DDDR pacemakers at lower rate behave similarly to DDD pacemakers when they operate on a ventricular timing concept. When a ventricular-sensed event (such as a PVC) occurs during the pacemaker VA interval, ventricular timed DDDR pacemakers will reset the VAI. Consequently, the

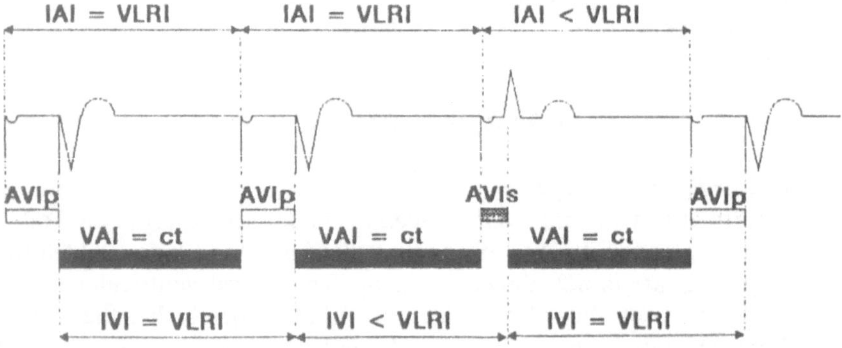

Figure 1. Ventricular timed DDD and DDDR pacing during lower rate operation. AVIp: programmed AV interval; AVIs: AV interval terminated by a sensed ventricular event; IAI: interatrial interval; IVI: interventricular interval; VAI: ventriculoatrial interval; VLRI: ventricular lower rate interval; ct: constant.

interval between that ventricular event and the subsequent ventricular stimulus will equal the ventricular lower rate interval (VLRI) (Figure 2).

(b) *Atrial timing concept*

However, there are also DDDR pacemakers which function on an atrial

Table 1. DDD and DDDR pacing at lower rate — ventricular timing (VLRI is fixed). VAI: ventriculoatrial interval; VLRI: ventricular lower rate interval; AVIp: programmed AV interval; IVI: interventricular interval; AVI_2: AV interval following the measured VAI; AVIs: AV interval terminated by a sensed ventricular event.

$$\text{VAI} = \text{VLRI} - \text{AVIp} = \text{constant}$$

$$\text{IVI} = \text{VAI} + \text{AVI}_2$$

$$\text{IVI} = \text{VLRI} - \text{AVIp} + \text{AVI}_2$$

* spontaneous AV conduction:
  $AVI_2 = AVIs < AVIp$
  then $IVI < VLRI$
* no spontaneous AV conduction:
  $AVI_2 = AVIp$
  then $IVI = VLRI$

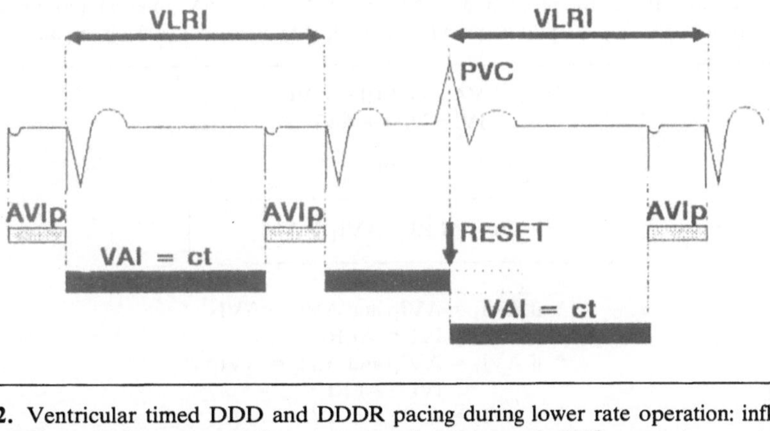

Figure 2. Ventricular timed DDD and DDDR pacing during lower rate operation: influence of a sensed ventricular event. AVIp: programmed AV interval; PVC: premature ventricular complex; VAI: ventriculoatrial interval; VLRI: ventricular lower rate interval; ct: constant.

timing concept. They differ from the previous generation of DDD pacemakers in that their basic pacing interval, like that of a normal heart, begins with an atrial event (A-A timing) rather than a ventricular event (V-V timing). In those pacemakers, the atrial lower rate interval (ALRI) is kept constant and is defined as being the longest interval between two consecutive atrial stimuli or between a sensed P wave and the subsequent atrial stimulus. Consequently, the pacemaker ventriculoatrial interval (VAI) can no longer be considered as the hallmark of normal pacing in DDD or DDDR pacemakers with atrial timing. Indeed, atrial timed DDD(R) pacemakers at their lower rate will have a pacemaker VA interval that equals the atrial lower rate interval (ALRI) minus the preceding AV interval (AVI_1) [Table 2]. The AV interval (AVI) is variable and may range between zero and the programmed AV interval (AVIp), since the programmed AV interval may be ended prematurely either by a spontaneous ventricular event or by an early ventricular stimulus during ventricular safety pacing. It follows that the interventricular interval (IVI) may be equal, shorter, but also longer than the programmed atrial lower rate interval (ALRI) (Figure 3).

Therefore, the normal operation of an atrial-timed DDDR pacemaker at lower rate should no longer be evaluated by the interventricular interval IVI (the interval between two consecutive ventricular events) but by the interatrial interval IAI (the interval between two atrial events). During atrial timed DDDR pacing at the lower rate, the interatrial interval (IAI) is always constant and equals the atrial lower rate interval (ALRI).

However, when special modes such as 'hysteresis' or 'rate drop at night' are programmed, the interatrial interval (IAI) can be longer than the pro-

Table 2. DDDR pacing at lower rate — atrial timing (ALRI is fixed).
ALRI: atrial lower rate interval; AVI_1: AV interval preceding the measured interventricular interval; AV_2: AV interval prior to the ventricular event terminating the measured interventricular interval; AVIp: programmed AV interval; AVIs: AV interval terminated by a sensed ventricular event; IVI: interventricular interval; VAI: ventriculoatrial interval.

$$VAI = ALRI - AVI_1$$
$$IVI = VAI + AVI_2$$

or

$$\boxed{IVI = ALRI - AVI_1 + AVI_2}$$

* if $AVI_1 = AVIp$ and $AVI_2 = AVIs$
  $IVI < ALRI$
* if $AVI_1 = AVIp$ and $AVI_2 = AVIp$
  $IVI = ALRI$
* if $AVI_1 = AVIs$ and $AVI_2 = AVIp$
  $IVI > ALRI$

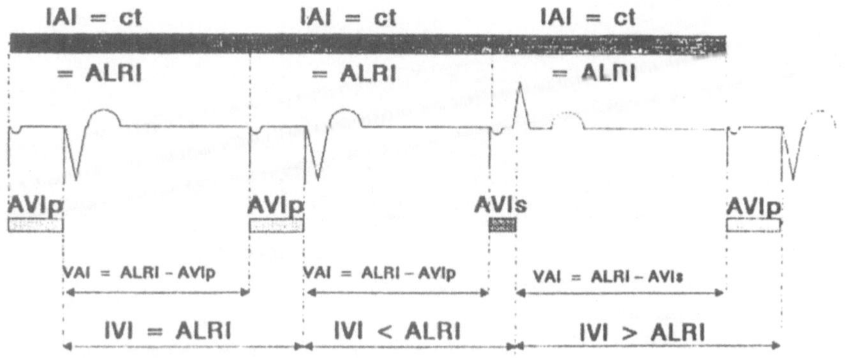

Figure 3. Atrial timed DDDR pacing during lower rate operation. ALRI: atrial lower rate interval; AVIp: programmed AV interval; AVIs: AV interval terminated by a sensed ventricular event; IAI: interatrial interval; IVI: interventricular interval; VAI: ventriculoatrial interval; ct: constant.

grammed atrial lower rate interval (ALRI). In the absence of those special modes, an increase of the interatrial interval (IAI) beyond the atrial lower rate interval (ALRI) is abnormal and may be due to false inhibition or impending battery depletion. When a ventricular sensed event (e.g. a PVC) occurs during the pacemaker ventriculoatrial interval, atrial timed DDDR pacemakers may behave differently according to the manufacturer. The interval subsequent to the sensed ventricular event may either be equal to the VAI (ALRI – AVIp) or equal to the atrial lower rate interval (ALRI) (Figure 4).

TRACKING

In the absence of a spontaneous atrial rhythm, the DDD pacemaker will stimulate both atrium and ventricle at the lower rate. When the intrinsic atrial rhythm is faster than the lower rate limit, a DDD pacemaker will synchronize the ventricle to the atrium in a 1:1 tracking mode as long as the upper rate limit (URL) or maximum ventricular tracking rate (MVTR) is not exceeded. However, when there is chronotropic incompetence, resulting in an inadequate rise of the atrial rate as a function of metabolic demand, the resulting ventricular pacing rate will be too low for the imposed workload.

DDDR pacemakers behave exactly in the same way as long as there is a sufficient spontaneous atrial rhythm. When the intrinsic atrial rate fails to increase sufficiently, the sensor will serve as a nonatrial marker of global

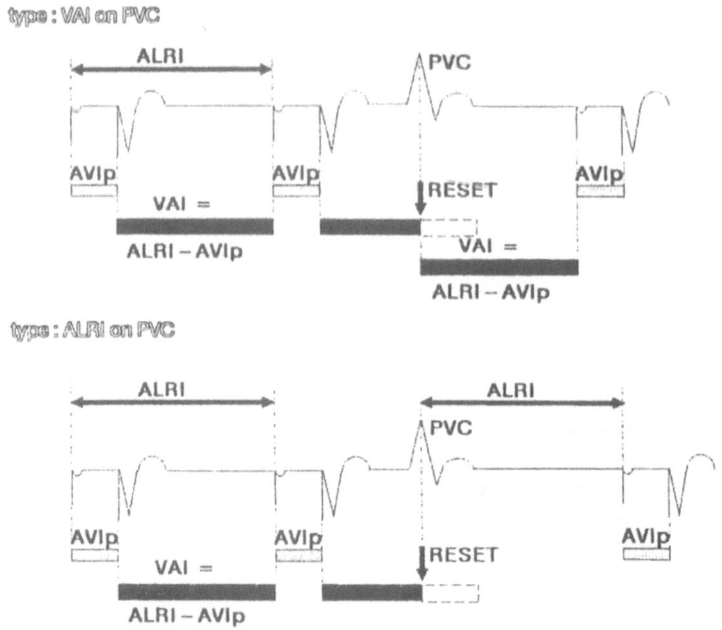

Figure 4. Atrial timed DDDR pacing during lower rate operation: influence of a sensed ventricular event. ALRI: atrial lower rate interval; AVIp: programmed AV interval; PVC: premature ventricular complex; VAI: ventriculoatrial interval.

metabolic demand and might be used as a determinant of physiologic pacing rate.

In DDDR pacemakers with *ventricular timing*, the ventriculoatrial interval (VAI) will be constant for a given input to the sensor and equals the sensor-indicated interval (SII) minus the AV interval (AVI) (Figure 5). If there is no intrinsic R wave, the interventricular interval (IVI) will be equal to the sensor-indicated interval (SII). However, if antegrade conduction is intact, the spontaneous AV interval (AVIs) might be shorter than the programmed one (AVIp), resulting in an interventricular interval which will be shorter than the sensor-indicated interval (SII). If there is alternation between paced and conducted ventricular events, the resulting mean pacing rate will therefore be slightly faster than the rate imposed by the sensor.

In *atrial-timed* DDDR pacemakers, the sensor will drive the atria at a constant rate, according to the metabolic demand, as long as the workload remains constant. In contrast to ventricular timing, the interventricular interval (IVI) may not only be shorter or equal to the sensor-indicated interval (SII), but also longer. Therefore the instantaneous ventricular rate can be faster or equal to but also lower than the sensor-indicated rate.

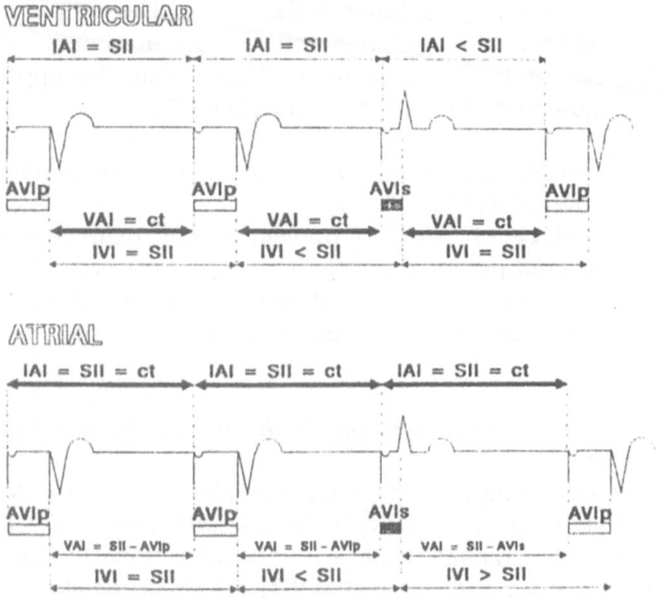

Figure 5. Ventricular versus atrial timing during sensor driven DDDR pacing. AVIp: programmed AV interval; AVIs: AV interval terminated by a sensed ventricular event; IAI: interatrial interval; IVI: interventricular interval; SII: sensor indicated interval; VAI: ventriculo-atrial interval; ct: constant.

To mimic the behavior of the normal AV node during physiologic sinus acceleration, some pacemakers are equipped with a rate-responsive AV interval. This may result in a progressive shortening of the AV delay as the sinus or sensor-driven rate increases.

UPPER RATE PACING

The behavior of a DDDR pacemaker at its upper rate depends upon the settings of the upper rate limit (URL) and the maximum sensor-driven rate (MSDR).

The upper rate limit (URL) *or maximum ventricular tracking rate* (MVTR) defines the maximum rate at which the ventricular output will be tracked in a 1:1 fashion to the spontaneous atrial rate (SAR). The upper value to which the upper rate limit (URL) can be programmed is restricted by TARI (sum of the AV interval (AVI) and postventricular atrial refractory interval (ARI) which determines the blocking onset rate (BOR) [1], i.e. the rate at which pacemaker AV block occurs.

The maximum sensor driven rate (MSDR) is the programmed highest atrial pacing rate driven by the input to the sensor. Theoretically, a selection can be made between three different settings: the maximum sensor driven rate (MSDR) can be lower, equal to, or higher than the upper rate limit (URL) or maximum ventricular tracking rate (MVTR).

(a) *Maximum sensor driven rate lower than the maximum ventricular tracking rate: MSDR < MVTR*
This option could be of value to prevent higher than clinically desired sensor-mediated atrial paced rates, e.g. in avoiding excessive high rates due to mechanical excitation of the piezoelectric sensor in patients likely to be exposed to heavy bodily vibration (air travel by helicopter, truck driving, etc.).

(b) *Maximum sensor driven rate equals the maximum ventricular tracking rate: MSDR = MVTR*
At atrial rates exceeding the maximum ventricular tracking rate and lower than the blocking onset rate (BOR), a Wenckebach response will occur in a DDD pacemaker programmed with a total atrial refractory interval (TARI) shorter than the upper rate interval (URI) [1]. A pacemaker Wenckebach cycle is characterized by a progressive prolongation of the interval between a sensed P wave and the successive ventricular stimulus until a P wave falling within the postventricular atrial refractory interval (ARI), creates a pause (Figure 6). To match the shorter interval between two consecutive P waves (SAI) to the longer ventricular upper rate interval (URI), an extension interval (AVE) is added to the programmed AV interval (AVIp).

The upper rate behavior of DDDR pacemakers may be different from that of DDD pacemakers as the pause during the Wenckebach cycle can be shortened by sensor-driven AV sequential pacing, i.e. sensor-driven rate smoothing [2].

Instead of creating a pause, the *ventricular timed* DDDR pacemaker may initiate an AV sequential sensor driven pacing (Figure 7) after a ventriculo-atrial interval (VAI) determined by the sensor driven interval (SDI) minus the programmed atrioventricular interval (AVIp):

$$\text{VAI} = \text{SDI} - \text{AVIp}.$$

If the maximum sensor-driven rate (MSDR) equals the maximum ventricular tracking rate (MVTR), the minimum value of the sensor-driven interval (mSDI) will equal the upper rate interval (URI).

In an *atrial-timed* DDDR pacemaker, the sensor-driven interval is re-started by each sensed P wave. However, since the ventricular upper rate interval (URI) may not be violated, the sensor driven interval (SDI) will also be reset by a ventricular paced event following a spontaneous P wave with a delay longer than the programmed AV interval (AVIp), i.e. whenever there exists an AV extension (AVE). Therefore, the last sensed P wave within the

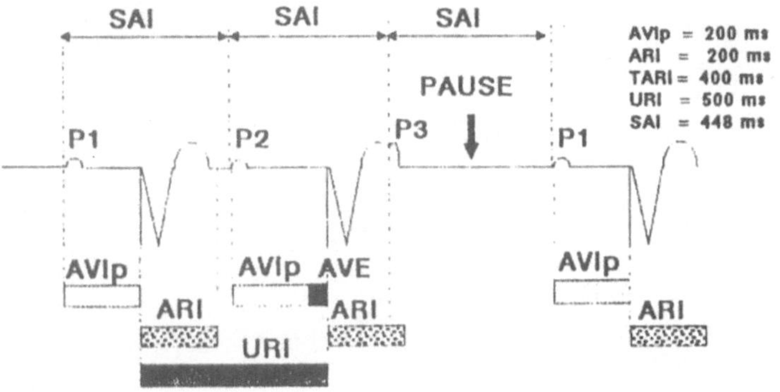

Figure 6. Wenckebach behavior in DDD pacemaker. ARI: postventricular atrial refractory interval; AVE: AV interval extension; AVIp: programmed AV interval; P1, P2: sensed P waves; P3: not sensed P wave; SAI: spontaneous atrial interval; TARI: total atrial refractory interval; URI: upper rate interval.

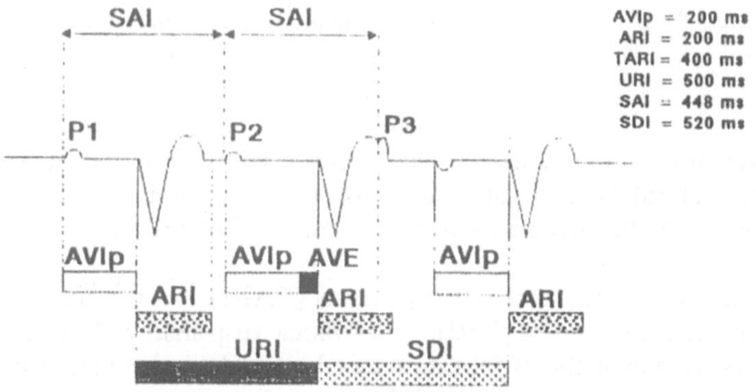

Figure 7. Wenckebach behavior in ventricular timed DDDR pacemaker. ARI: postventricular atrial refractory interval; AVE: AV interval extension; AVIp: programmed AV interval; P1, P2: sensed P waves; P3: not sensed P wave; SAI: spontaneous atrial interval; SDI: sensor driven interval; TARI: total atrial refractory interval; URI: upper rate interval.

Wenckebach cycle will initiate a sensor-driven interval (SDI) which will be reset by the consecutive paced ventricular event. When the newly started sensor-driven interval (SDI) is timed out, an atrial pacing pulse will be delivered, resulting in a shortening of the pause (Figure 8).

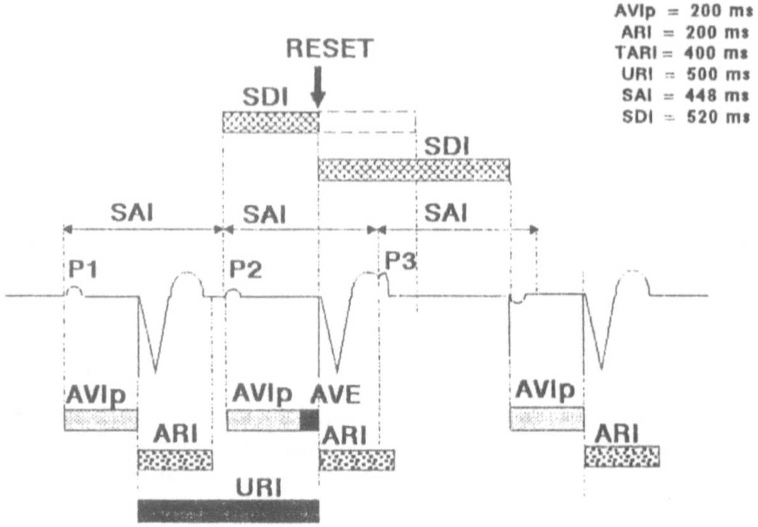

Figure 8. Wenckebach behavior in atrial timed DDDR pacemaker. ARI: postventricular atrial refractory interval; AVE: AV interval extension; AVIp: programmed AV interval; P1, P2: sensed P waves; P3: not sensed P wave; SAI: spontaneous atrial interval; SDI: sensor driven interval; TARI: total atrial refractory interval; URI: upper rate interval;

However, if a spontaneous P wave is sensed before the end of the sensor-driven interval (SDI), the pause will not be shortened, i.e. it will remain identical to the pause occurring in the normal DDD pacemaker (see Figure 6).

When the total atrial refractory interval (TARI) is programmed shorter than the upper rate interval (URI), a 2:1 block response will occur in DDD pacemakers whenever the atrial rate exceeds the blocking onset rate (BOR). At optimal programming of the DDDR pacemaker, the pauses that would be created by the block response will be replaced by AV sequential pacing according to the sensor input. The maximum sensor-driven rate (MSDR) may be equal or less than the blocking onset rate (BOR), i.e. the minimum sensor-driven interval (mSDI) has to be larger or at least equal to the total atrial refractory interval (TARI).

When the total atrial refractory interval (TARI) is programmed equal to the upper rate interval (URI), DDD pacemakers will show a direct transition from 1:1 atrial tracking to 2:1 block response. DDDR pacemakers will have an identical behavior, as explained earlier for atrial rates exceeding the blocking onset rate (BOR). Optimal programming of the rate-response parameters of DDDR pacemakers may avoid the wide swings in ventricular cycle lengths that could occur in DDD pacemakers during the transition from 1:1 atrial tracking to Wenckebach behavior or block response [3].

(c) *Maximum sensor driven rate (MSDR) greater than maximum ventricular tracking rate (MVTR): MSDR > MVTR*
A discordant programming of the maximum sensor driven rate (MSDR) may be of clinical interest, especially in patients prone to developing atrial tachyarrhythmias or in patients at risk from pacemaker mediated endless loop tachycardias. Limiting the maximum ventricular rate (MVTR) to a low value, safeguards the patient against excessively high ventricular rates during 1:1 tracking at the occurrence of an atrial tachyarrhythmia. Moreover, it will limit the ventricular rate during an artificial endless loop tachycardia. While protecting the patient against unwanted excessively high ventricular pacing rates, a low maximum ventricular tracking rate (MVTR) also restricts the maximum ventricular rate during exercise in active patients. Therefore the maximum sensor-driven rate (MSDR) should be programmed above the maximum ventricular tracking rate (MVTR) in those patients in order to maintain AV sequential pacing at higher rates.

CONCLUSION

Dual chamber rate-responsive pacemakers have a distinct advantage over the DDD mode as they can provide rate responsive pacing in individuals whose sinus node is not suitable as an indicator of physiologic heart-rate modulation. Moreover, DDDR pacing protects patients from precipitous drops in pacing rate which may result from unsensed P waves during Wenckebach behavior and 2:1 block at the upper rate limit. As the rate of DDDR devices is determined by the combined input from the intrinsic atrial rate and the sensor driven rate, ECG interpretation may become more complex. Therefore, the clinician should be familiar with the basic concepts and functions of such pulse generators to avoid misdiagnosing of pacemaker malfunction.

REFERENCES

1. Stroobandt R, Willems R, Holvoet G, Backers J, Sinnaeve A (1986) Prediction of Wenckebach behavior and block response in DDD pacemakers. *PACE* 9: 1040—1046
2. Higano S, Hayes D (1989) P wave tracking above the maximum tracking rate in a DDDR pacemaker. *PACE* 1044—1048
3. Higano S, Hayes D, Eisinger G (1989) Sensor-driven rate smoothing in a DDDR pacemaker. *PACE* 922—929

8. Facing the 'faces' of pacing after implantation

MARC GOETHALS, WILLY TIMMERMANS, ROGER WILLEMS, ERIK ANDRIES & ROLAND STROOBANDT

INTRODUCTION

The use of long-lived lithium batteries and hermetically sealed circuitry have much increased the longevity and reliability of the presently used pacemaker systems, such that the detection of premature power source depletion and pacemaker system malfunction is less of a problem than in the past. Through the addition of multiprogrammability, dual chamber pacing and rate responsive behaviour the interaction between the patient and the pacing system has become much more complex.

1. ORGANIZATION OF A PACEMAKER FOLLOW UP CLINIC

The main objectives of a pacemaker clinic in the nineties can be defined as:

1. to predict impending pacemaker system failure before the patient is at risk;
2. to determine the nature of (impending?) pacemaker system failure and to correct it by noninvasive ways whenever possible;
3. to maximize pacing system longevity by determination of stimulation thresholds at regular intervals and appropriately adjusting the pacemaker output;
4. to optimize pacemaker system performance by the judicious use of programmability according to the patients needs;
5. to record patient location if a recall occurs due to some pattern of systematic failure of lead or generator;
6. to minimize medical complications from the implanted system (e.g. skin erosion, pacemaker mediated tachycardia, etc.);
7. to aid in the generation of statistical data on an international or nation wide basis in order to make early recognition of systematic failure of pacemaker generators or leads possible.

E. Andries, P. Brugada & R. Stroobandt (eds.), How to face 'the faces' of cardiac pacing, 139—181.

1.1. Equipment

In the pacemaker follow-up clinic, the following equipment should be routinely available:

1. an analog multichannel electrocardiograph;
2. a digital pulse counter for measurement of pulse duration and interval;
3. a magnet;
4. various programmers appropriate to the various pacemaker models that are in use in the hospital. Contact addresses and telephone numbers of the other pacemaker companies should be available;
5. an external pacemaker (preferably with extrastimulus and high rate facilities) with connecting cables and skin electrodes for chest wall stimulation;
6. eventually an oscilloscope with isolation amplifier and polaroid camera or alternative equipment for high fidelity recording of impulse wave forms;
7. pacemaker records to be kept separate from the patient record system or alternatively a computer data base. Additionally, there should be easy access to X-ray and echocardiography.

Equipment

1. electrocardiograph
2. digital pulse counter
3. magnet
4. programmers
5. external pacemaker for chest wall stimulation
6. oscilloscope with isolation amplifier and camera
7. pacemaker records

1.2. Organization of follow-up visits

The provision of general medical or cardiologic care in the pacer clinic has often been criticized [1], but in Europe it is certainly common practice. Moreover, the presence of a conduction disturbance severe enough to warrant the implantation of a cardiac pacemaker often betrays a general disorder of cardiac function. It is hard, if not impossible, to separate medical care for the patient from pacemaker programmability, which permits optimal adjustment of pacemaker function to the need of the patient.

Patients should be seen on a regular basis and not solely at the request of the patient or the family physician because of persistence or return of symptoms. The schedule should be based on what is known about threshold behaviour after electrode implantation. In view of the sometimes pronounced threshold rises in the first few weeks after electrode insertion, the first pacemaker follow-up visit is scheduled at about 6 weeks post implantation and every six months thereafter.

1.3. Pacemaker records

The pacemaker records contain both patient and pacemaker data.

1. *Patient data* consists of identification of the patient, including address and phone number, indication for pacemaker implantation, an evaluation of pacemaker dependency, a report of the implantation, and eventual pacemaker revisions.

The implantation report contains information on the venous access route(s), peculiarities encountered during the implantation (e.g. persistent left superior vena cava, absence of the right atrial appendage due to previous cardiac surgery, etc.), stimulation threshold measurements during implantation, measurement of lead impedance, unfiltered registration of the intra-atrial and/or intra-ventricular electrocardiogram, measurement of the magnitude and slew rate of the intra-cardiac signal with a pace-system analyser, and exact location of the electrodes within the heart chambers.

On each follow-up visit, the electrocardiogram and the pacemaker dependency must be reevaluated and updated in the record.

2. *Pacemaker data* consists of

* model and serial number of pulse generator and lead(s);
* pacemaker program settings;
* end-of-life indication of the pulse generator;
* pacing and sensitivity threshold measurements;
* if available, telemetry data on lead impedance, battery status, and intra-cardiac electrocardiogram.

1.4 Pacemaker follow-up evaluation

The following information is obtained at regular intervals:

1.4.1. An *X-ray* is obtained immediately after implantation and once a year thereafter, except when pacemaker malfunction is noted as suggestive of lead dislodgement, perforation, or fracture. Low penetration X-rays may be useful when polyurethane degradation is suspected.

1.4.2. A *twelve lead electrocardiogram* with and without *magnet* application, including the transitional phase after application of the magnet, is registered on each visit. If the electrocardiogram displays only paced rhythm, it is wise to make a registration of the *temporary inhibition* of the pacemaker. Facilities for temporary inhibition are often provided on the programmer, if not, inhibition of unipolar pacemakers can easily be achieved by chest-wall stimulation. Inhibition of the pacemaker allows evaluation of pacemaker-dependency.

1.4.3. *Stimulation thresholds* are obtained on each visit in order to optimize pacemaker longevity and, at the same time, provide an adequate safety margin to the patient.

The stimulation threshold is defined as the smallest stimulus that consistently depolarizes the myocardium when applied outside the refractory period of the heart.

Most of the presently available pacemaker systems are provided with an automated threshold measurement system with a stepwise reduction of impulse duration or amplitude.

If the threshold is measured by stepwise reduction of voltage or current at constant pulse width, for chronic lead implants adequate safety is provided to the patient if the output is programmed at twice the threshold value [2]. If, however, the threshold is determined by decreasing the pulse width at constant voltage or current, programming the impulse duration to twice the threshold value may not provide adequate safety. Therefore, the impulse width is programmed to three to four times the threshold value in order to provide adequate safety. Actually, programming the pulse width to very low values, not only provides inadequate safety to the patient but may increase energy consumption as well, as energy output decreases with decreasing pulse width down to a certain optimum pulse width and again increases with pulse width below the optimum value [3].

In the first few weeks after electrode insertion, it is prudent to leave the pacemaker output at its nominal settings in view of the sometimes pronounced transitory threshold rise due to inflammation around the electrode tip (electrode maturation). For atrial electrodes, this peaking of the stimulation threshold may occur earlier and its often more pronounced [4], sometimes with a loss of capture (or sensing) in the first few weeks after lead insertion. This constitutes, of course, less of a problem than the less-frequent pronounced threshold rise with ventricular lead implants.

1.4.4. *Sensing thresholds* are defined as the least sensitive settings of sensitivity (highest value) that consistently inhibits or triggers the appropriate pacemaker channel. In *single chamber* units, programming the unit too sensitive may cause inhibition of the pacemaker by electromagnetic interference or myopotentials. Conversely, choosing inappropriately low sensitivity may cause undersensing of the native complexes and sometimes induction of arrhythmias if a suitable arrhythmogenic substrate is present. The sensing threshold can be determined only when spontaneous activity is present, most conveniently by programming to the SST mode (AAT or VVT) and programming to progressively lower sensitivity (higher mV values). Transmission of endocardial electrograms can be of help but measurement of the amplitude of the transmitted signal does not always predict the sensing threshold, due to variable filtering of the signal by the amplifiers of the pacemaker and/or the pacemaker programmer.

In *dual chamber* units periodic evaluation of atrial sensing threshold is important.

In patients with atrioventricular conduction disturbances, this can easily be accomplished by programming the lower rate below the spontaneous sinus rate and progressively decreasing the atrial sensitivity up to the highest value that reproducibly allows the triggering of a ventricular stimulus by each P wave. In patients with intact atrioventricular conduction, the same procedure can be followed if the atrioventricular delay is abbreviated to a value well below the PR interval (with creation of a clearly identifiable ventricular pacemaker-evoked response) or, alternatively, the pacemaker can be programmed to the AAT mode.

Programming atrial sensitivity too low (high mV value) may cause loss of normal atrioventricular sequencing and can cause pacemaker mediated tachycardia, due to dissociation of atrial and ventricular depolarization [5]. Programming atrial sensitivity too high (low mV value) may cause oversensing of myopotentials by the atrial amplifier, resulting in fast ventricular pacing or dissociation of ventricular and atrial activation, which may again induce pacemaker-mediated circus movement tachycardia. It has to be remembered that, in the early postoperative days after insertion of an atrial lead, the endocardial P wave amplitude typically decreases, which may result in temporary malsensing [6]. Thereafter, an increase toward the acute value appears the rule, and malsensing generally resolves spontaneously [7—9].

Pectoral muscle exercises can be of help in the patient with symptoms suggestive of myopotential inhibition or pacemaker mediated tachycardia, or simply as a tool to elicit pacemaker mediated tachycardia.

1.4.5. Determination of *pulse duration* and *interval.* The pulse duration and interval are determined with a digital pulse counter.

The pulse interval is the time in milliseconds between two consecutive pulses and is measured with and without application of a magnet. The pulse interval with or without a magnet contains the *end-of-life indication* for most pacemakers. At present, all pulse generators decrease the rate or increase the pulse interval with battery depletion, but the end-of-life criteria vary among different pacemaker manufacturers and models of generators.

Rate decrease can be progressive or stepwise (e.g. from 72 bpm to 62 bpm, the elective replacement indicator, to 52 bpm, the actual end-of-life indicator). In the early days, the end-of-life indication consisted of a parallel decrease of rate in both the magnet and the free-running mode. Some models provide stability of the rate in the free-running mode, whilst only the magnet rate decreases with battery depletion. Determination of rate or pulse interval is best done with the pacemaker programmed to its nominal settings.

In some models, a magnet response can only be elicited after explicit programming to the magnet mode.

The pulse duration is the time in milliseconds between the ascending and the descending limb of the pulse.

In some pacemakers, battery depletion is accompanied by an increase in pulse duration. The increase in pulse duration is intended to compensate for the decrease in pulse height and keeps the energy content of the pulse relatively constant in order to preserve effective capture within certain limits.

For some pacemakers, the increase in pulse duration can be used as a secondary end-of-life indicator.

1.4.6. Analysis of *pulse wave morphology.* The frequency content of the pacemaker impulse is beyond the frequency response of the ordinary electrocardiograph. The impulse can, however, be registered with an oscilloscope or can be reconstructed with a computer based system.

The analysis of the pulse-wave form can give some information about the integrity of the pacemaker and connected lead. Constant current pulse generators are not suitable for pulse-wave-form analysis but have virtually disappeared from the market.

All the presently commercialized pulse generators are of the constant voltage type. The pacemaker signals are fed into the oscilloscope by means of the classic Einthoven leads I, II, and III. The voltage that is detected between the poles of the Einthoven leads is proportional to the current flowing through the body between the poles of the pacemaker. As the dipole of bipolar pacemakers is very small, it follows that bipolar pacemakers are not suitable for pulse-wave form analysis. Pulse-wave-form analysis can only be performed reliably with unipolar systems.

The pulse wave is caused by the discharge of a condensor across the body and the lead. During the pulse, the current decreases exponentially according to a time constant $(R_i + R_1 + R_t) \times C$, with R_i being the internal resistance, R_1 the resistance of the lead, R_t the tissue resistance, and C the capacitance of the coupling condensor in the pacemaker. With the internal resistance R_i and the capacitance considered constant, it follows that the time constant is, to a large extent, influenced by the resistance of the lead R_1 and the tissue resistance R_t. The time constant of the exponential decay can be obtained by manually or electronically digitizing the impulse wave form [10, 11]. In order to obtain the time constant reliably, the pulse generator must be programmed to a long pulse duration.

With an *increase in tissue resistance* which could lead to exit block, the time constant increases and the slope of the exponential decay will thus decrease. With a *broken electrode* or *bad contact* between pacemaker and electrode, but *intact insulation* and *without a fluid layer* between the metal surfaces, similarly the increase in lead resistance R_1 is translated into a decrease of the slope of the impulse.

With *a broken electrode, intact insulation* and *a fluid layer between the broken ends*, the build up of a polarization voltage in the fluid layer causes a rapid drop of the current with a typical appearance of the pulse wave form.

With an *insulation leak*, the lead resistance decreases due to the creation of a parallel circuit, thus increasing the slope of the pulse-wave form.

With the availability of high-quality X-ray examination, telemetry of the endocardial electrogram and of lead impedance, analysis of pulse-wave morphology is rarely an essential link in the establishment of a correct diagnosis in the patient with pacemaker malfunction.

1.4.7. Evaluation of *pacemaker-dependency*. For practical purposes, the patient is pacemaker-dependent if the sudden loss of pacing will result in a Stokes—Adams episode, serious injury, or death [12]. Pacemaker-dependency can vary from time to time and it is obvious that the worst score a patient ever has defines his status. In our follow-up clinic, therefore, pacemaker-output is briefly inhibited on each visit. If inhibition of pacemaker-output with chest-wall stimulation or through the programmer for four seconds result in asystole without an escape rhythm, the patient is considered to be completely pacemaker-dependent [12]. The evaluation of pacemaker-dependency can be of considerable help in the follow-up of the patient. In pacemaker-dependent patients, generator replacement is advised at an earlier stage when the unit approaches its end-of-life indication. Similarly, pacemaker revision is performed sooner in those patients, when information becomes available on systematic failure of certain types of generator or lead.

1.4.8. Adjustment of *programmable functions*. The availability of multiprogrammable generators allows fine tuning of the pacemaker function to the needs of the individual patient. The problem of multiprogrammability is dealt with below.

Follow-up check list
1. X-ray (annually)
2. ECG ± magnet and with PM inhibition
3. stimulation threshold
4. sensing threshold / pectoral muscle exercise
5. pulse duration and interval / end-of-life indication
6. pulse-wave morphology
7. pacemaker dependency
8. programmable functions

2. THE PACEMAKER ELECTROCARDIOGRAM

A 12-lead electrocardiogram with and without magnet application, including the transitional phase after application of the magnet, is obtained.

The much more complex issue of dual chamber pacemaker-electrocardiography is dealt with in a separate chapter. The following discussion is limited to the interpretation of ECG in single chamber units. The first step in the interpretation of the ECG consists of identifying the pacemaker spikes (Stimulus artefact) with two obvious possibilities:

2.1. No stimulus artefact can be identified

This means either malfunction of the pacemaker or inhibition by electromagnetic interference or spontaneous electrical activity, which should be faster than the programmed rate of the pacemaker, except when a negative hysteresis is programmed, which allows the spontaneous rhythm to decrease below the programmed rate of the pacemaker.

Pacemaker spikes can be elicited by application of a magnet, which converts most units to the asynchronous mode (SOO) and allows the evaluation of pacemaker capture, i.e. the ability of the pulse to bring about depolarization of the paced chamber.

One should take into account, however, that due to electromagnetic interference, the response of the pacemaker may be irregular in the transitional phase and that moving the magnet up and down, so-called 'magnet waving' [13], may even inhibit the pacemaker. Moreover, in some pacemakers, the magnet response may be temporarily or permanently programmable, i.e. the Pacesetter AFP 283 automatically reverts to the 'magnet off' mode after interrogation of telemetric data, after which the unit must be explicitly programmed to the 'magnet on' mode in order for asynchronous pacing to occur on application of a magnet [14].

In addition, some pulse generators perform a threshold margin test on application of a magnet. The Medtronic Spectrax and Minix SSI units revert to a rate of 100 bpm for three consecutive intervals (600 ms) after magnet application. At the end of the third 600 ms interval, the pulse width is reduced to 75 % of the programmed width. Thereafter, asynchronous pacing at the programmed rate ensues until the magnet is removed (Figure 1).

The Siemens pacemakers and some of the Telectronics models automatically perform a threshold test on magnet application when the Vario function is programmed 'on'; the output of the pacemaker is decreased by 1/16 with each successive pulse, the sequence ending with a marker pulse (Figure 2). The pacing threshold can be adequately determined if the programmed pacer output is known. The sequence is automatically interrupted when the magnet is removed.

2.2. A stimulus artefact can be identified on the ECG

A systematic approach is warranted and includes a careful analysis of the pacemaker spike, its relation with the native complexes and the characteristics of the pacemaker-evoked response.

2.2.1. *The pacemaker stimulus artefact*

Spikes of unipolar pulse generators are of high amplitude compared to the native complexes and are often distorted or attenuated by the filters of a conventional analog electrocardiograph. Spikes of bipolar pulse generators, however, are often difficult to identify.

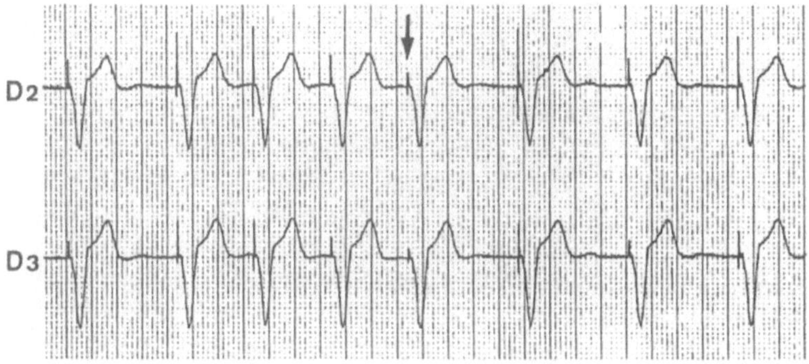

Figure 1. Safety margin test in a Medtronic Spectrax VVI pacemaker.
Lead D2 and D3 in a patient with a VVI Medtronic Spectrax pacemaker. After the first beat from the left, the magnet is applied eliciting pacing at a frequency of 100 beats per minute (cycle length 600 ms) for three successive cycles. Thereafter, the frequency returns to the programmed value. the fourth pacemaker spike, ending the third 600 ms cycle (arrow) clearly has a lower amplitude.

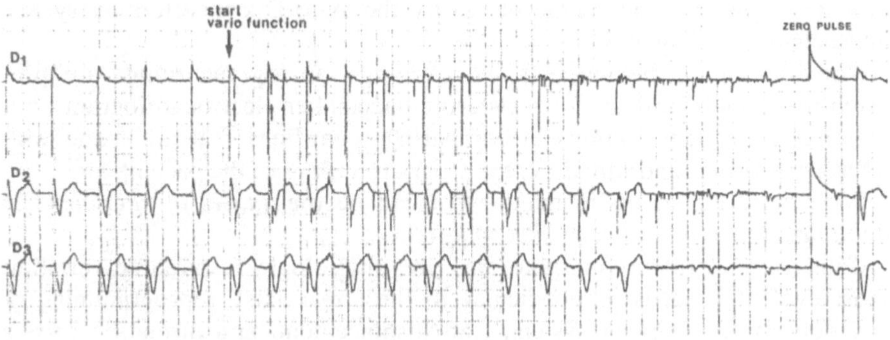

Figure 2. Vario test in a Siemens VVI pacemaker.
Three standard ECG leads are shown. The large arrow indicates activation of the vario function. Pacemaker output is stepwise decreased by 1/16 of its programmed value up to the zero pulse with reversed polarity. Thereafter pacing at the programmed settings resumes. During the vario test, random electrical signals are commonly observed (small arrows) due to the discharge of the output capacitor.

Nowadays, pacemakers are provided with a rapid recharge pulse in order to diminish polarization (Figure 3). The rapid recharge pulse is opposite in polarity to the main stimulus artifact. Both pulses influence the analog electrocardiograph which may display both pulses accurately or, alterna-

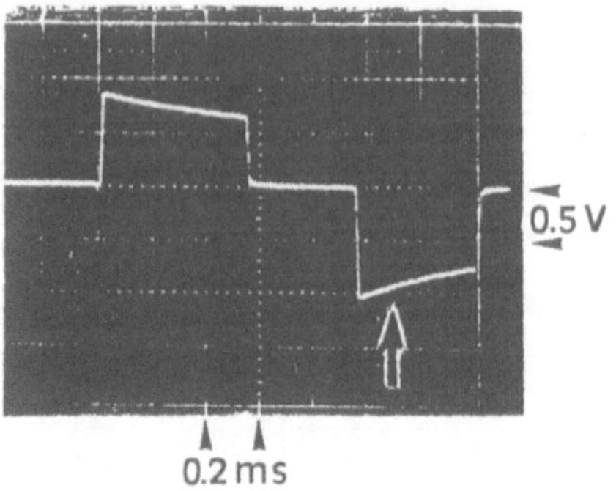

Figure 3. The rapid recharge pulse.
Photoanalysis of a pacemaker pulse demonstrating the fast recharge pulse with reversed polarity (see arrow).

tively, a mixture of both. The relative blend of these pulses determines the display of the combined artefact, which may appear as a positive, a negative, or a biphasic form [15]. With digital electrocardiographs, the stimulus artefact may show varying amplitudes on a beat-to-beat basis due to the intermittent sampling of the ECG voltage by the A-to-D converters every few milliseconds [16] (Figure 4).

Notwithstanding the inherent shortcomings, analog recordings are preferred to digital recordings for registering pacemaker electrocardiograms, but one should be very cautious in comparing recordings that are made with different recorders and with different program settings of the pacemaker.

With these limitations in mind, what can be the *causes of alterations in amplitude or axis of the pacemaker spike*?

Causes of alterations in spike axis are physiologic respiratory variation (most evident in leads perpendicular to the spike axis), reprogramming of lead polarity in a bipolar system, lead displacement, and insulation breaks with electrical continuity of the electric conductor [17].

Only gross alterations in spike amplitude can be considered significant and may be due to pulse-generator failure or alteration in the programming of the pulse generator, insulation break either internally (within a bipolar lead) or externally, partial wire fracture with fluid bridge, and different filtering characteristics of electrocardiographs.

2.2.2. *Characteristics of the pacemaker evoked response*

With bipolar pacemakers, the stimulus artefact is generally small but with unipolar pacemakers, the stimulus artefact is much larger and the onset of the P wave or QRS complex may be heavily distorted.

The deformation may result from technical shortcomings of the electrocar-

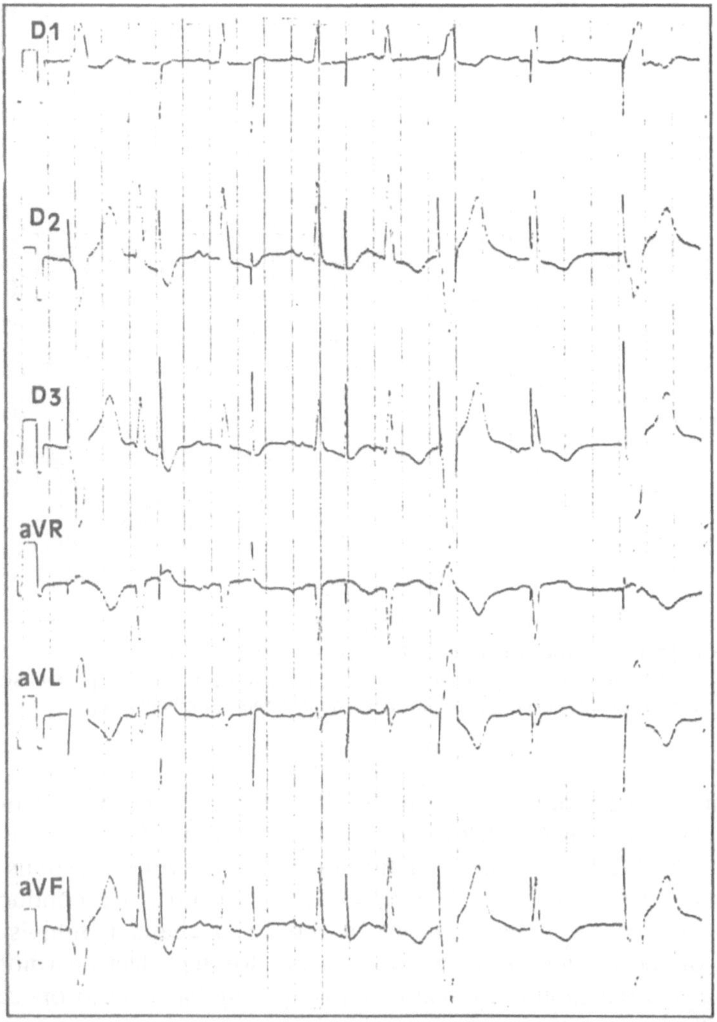

Figure 4. Distortion of the pacemaker stimulus artefact by digital electrocardiographs. Limb leads in a patient with a VVI pacemaker. The pacemaker functions temporarily in the VOO mode due to magnet application. The stimulus artefact varies widely not only in amplitude but also in frontal plane axis (courtesy of Dr I. Kersschot).

diograph or alternatively from polarization voltages built up at the electrode tissue interface [18].

With atrial or atrioventricular sequential pacing the recognition of atrial capture (Figure 5) may be difficult especially with unipolar systems. With

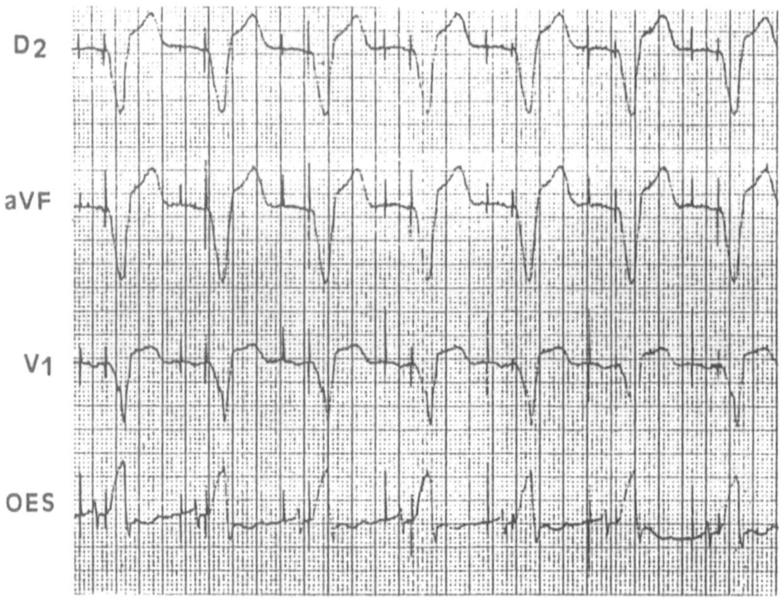

Figure 5. Atrial capture during pacing from the right atrial appendage.
Three surface leads and an oesophageal lead in a patient with a DDD pacemaker. Atrial capture can be seen in lead V1 but is much more evident in the oesophageal lead.

atrial pacing, atrial capture can be indirectly inferred from the ensuing QRS complex. Similarly, the AV delay can be lengthened in AV sequential pacers in order to facilitate recognition of the atrial-paced evoked response or to infer it from the ensuing QRS complex, if atrioventricular conduction is intact. Decreasing the voltage output of the atrial channel diminishes the distortion of the P wave. A complete strategy for the electrocardiographic delineation of atrial capture in dual-chamber pacing has been proposed [19]. Occasionally, one has to resort to nonelectrocardiographic means in order to demonstrate atrial capture such as echocardiography or cardiac Doppler [20, 21]. Atrial capture can be reliably demonstrated with subcostal right atrial free-wall M-mode echocardiography or from the presence of atrial waves in the M-mode or Doppler pattern of the mitral or tricuspid valve.

The electrocardiographic characteristics of the *atrial-paced evoked response* obviously depend on the location of the electrode and pulse generator. The polarization artefact with unipolar pacemakers has decreased with the introduction of the fast recharge pulse and is much less pronounced in bipolar pacemakers. With the classical position of the pacemaker in the right pectoral region and the electrode in the right atrial appendage, the atrial depolarization wave is most distorted in lead II and least in lead III and aVL by polarization artefact.

From the right atrial appendage, the atrial-evoked response is prolonged and of diminished amplitude compared to sinus rhythm. The frontal plane vector of the complex is directed, inferiorly initially, and then shifts leftward and superiorly [22].

The configuration of the *ventricular-paced evoked response* depends on the site of stimulation [23]. Pacemaker wires attached to the left ventricle produce a right bundle branch-block configuration, whereas wires attached to the right ventricle, either epi- or endocardially produce a left bundle branch-block pattern. With pacing from the *right ventricular apex*, the left ventricle is activated in a retrograde manner (from apex to base) and the main QRS vector is to the left, superior and posterior. Therefore, on the ECG, a left bundle branch-block pattern, with superior and mostly leftward axis deviation, is inscribed and lead V6 is predominantly negative or biphasic (Figure 6).

Occasionally, the frontal-plane axis may be superior and to the right, yielding a positive complex in aVR, and a predominantly negative complex in I, II, III, and aVF.

The precordial leads in right-apical pacing thus show a left bundle branch pattern with negative complexes in V1—V2, but the complexes in V5—V6 may be either negative or positive.

The pattern of *right ventricular-inflow* pacing (near the tricuspid valve) is very similar to that of right ventricular-apical pacing, but lead V6 tends to be predominantly positive (Figure 7).

With *right ventricular outflow-tract* pacing, such as may be the result of lead dislodgement, the frontal plane axis shifts to the normal range, i.e. inferiorly and lead V6 also becomes predominantly positive (Figure 8).

The presence of a *positive* complex in V1 should be considered abnormal [17] during transvenous pacing until proven otherwise, and may be a sign of an abnormal position of the lead in the coronary sinus (Figure 9) or in the left ventricle due to perforation (Figure 10).

If a pacemaker-stimulus outside the refractory period is observed without evoked response, *loss of capture* of the paced chamber is present due to [24]

1. micro- or macroscopic dislodgement or perforation of the electrode
2. inapparent placement of the electrode in the cardiac venous system
3. mismatch between the threshold for pacing and pacemaker output, i.e.
   - threshold rise with maturation of the electrode [25] which reaches a maximum in the first few weeks after implantation,
   - threshold rise due to electrolyte imbalance
     - infarction or ischemia
     - antiarrhythmic drugs
     - defibrillation

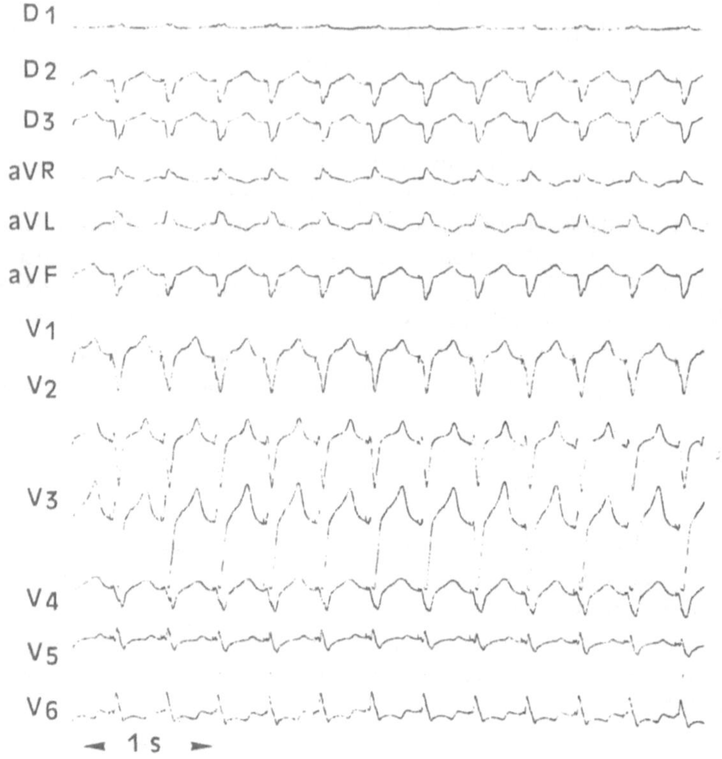

Figure 6. Ventricular pacing from the right ventricular apex.
Complete 12 lead ECG of a patient with a temporary pacemaker in the right ventricular apex. Note the left bundle branch block configuration with superior frontal axis, negative complexes in the right precordial leads, biphasic complexes in V6.

— inappropriate low programming of pacemaker-output parameters
— electrode problems: fracture, insulation break, or short circuit
— pulse-generator failure due to battery exhaustion or component failure, which may be induced by defibrillation or therapeutic radiation.

However, one should take into account that the pacemaker stimulus may obscure pacemaker-evoked response leading to apparent loss of capture by saturation of the amplifiers of some electrocardiographs or by automatic-gain compensation (Figure 11).

Apparent loss of capture occurs when the stimulus falls during the refractory period of native complexes due to undersensing. This is frequently the case with ventricular premature beats or with atrial fibrillation and fast ventricular response.

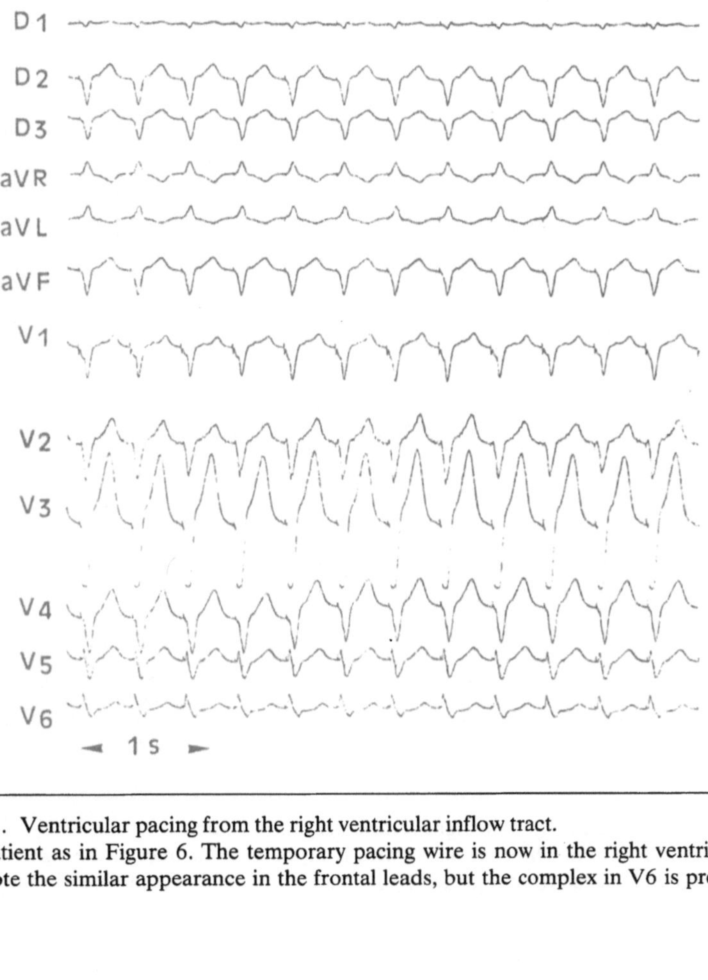

Figure 7. Ventricular pacing from the right ventricular inflow tract.
Same patient as in Figure 6. The temporary pacing wire is now in the right ventricular inflow tract. Note the similar appearance in the frontal leads, but the complex in V6 is predominantly positive.

2.2.3. *Relation of the pacemaker spike to the native complexes*
With normal sensing of the pulse generator, the time interval between the spontaneous complex and the following pacemaker stimulus equals the pacemaker automatic interval unless a hysteresis function has been programmed on.

2.2.3.1. *Undersensing.* If the interval between the spontaneous complex and the next stimulus artefact is shorter than the programmed pulse interval, the spontaneous activity does not reset the pacemaker and *undersensing* is present.

Undersensing of the native complexes may be due to [17]

1. inadequate amplitude of the cardiac signal due to malpositioning of the

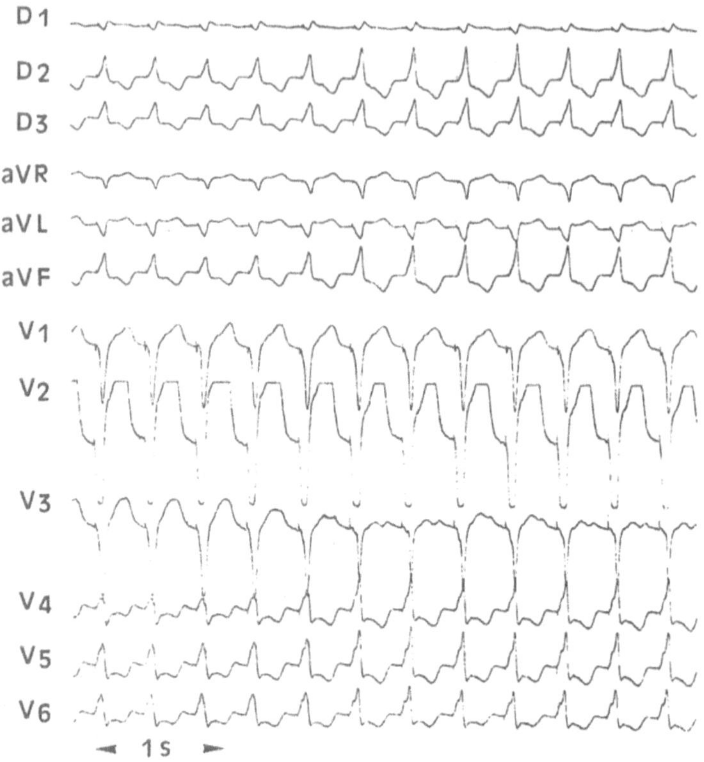

Figure 8. Ventricular pacing from the right ventricular outflow tract.
Same patient as in Figure 6. The pacemaker wire is in the outflow tract of the right ventricle. Note the inferior shift of the frontal plane QRS vector with positive complexes in the inferior leads.

electrode, micro- or macrodislodgement, perforation, myocardial infarction, chronic fibrosis and scarring around the electrode, electrolyte imbalance or antiarrhythmic drugs, and development of conduction disturbances such as bundle branch block.

On 24 hour Holter monitoring, 15% of patients with normal sensing of supraventricular complexes exhibit unsensed ventricular extrasystoles [32];

2. attenuation of an adequate cardiac signal upon entry in the pacing system:
   - mismatch between input and source impedance (e.g. a combination of a large surface area electrode with a low input impedance pulse generator) [26]
   - attenuation of the signal by an insulation leak or poor adaptor connectors;

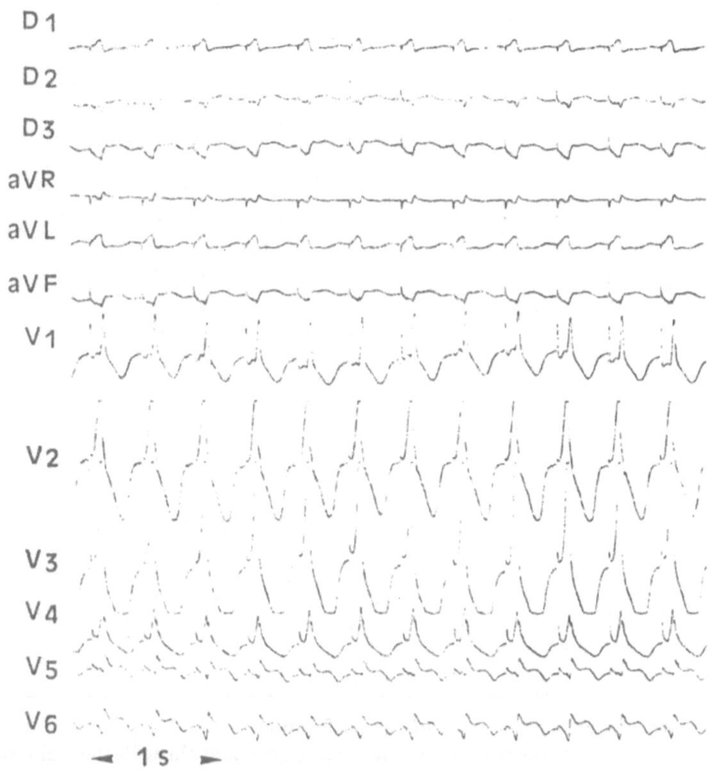

Figure 9. Ventricular pacing from the coronary sinus.
Same patient as in Figure 6.The pacemaker wire is in the proximal coronary sinus. Note the positive complex in the right precordial leads.

3. adequate cardiac signal delivered to a malfunctioning pulse generator, such as occurs with a reed switch remaining in the closed position [27].

2.2.3.2. *Apparent undersensing.* True undersensing must be differentiated from apparent undersensing. Two circumstances are discussed, i.e. the occurrence of pseudofusion beats and the interference reversion mode.

Pseudo-fusion beats consist of the superimposition of an ineffectual pacemaker spike upon a QRS complex originating from a single focus (i.e. the morphology of the pseudo-fusion beat is not intermediate between the native complex and the pacemaker-evoked response). Pseudo-fusion beats are manifestations of normal demand pacing and can be explained by the fact that a large portion of the surface QRS may be inscribed before the occurrence of the intrinsic deflection of the endocardial QRS, which actually resets

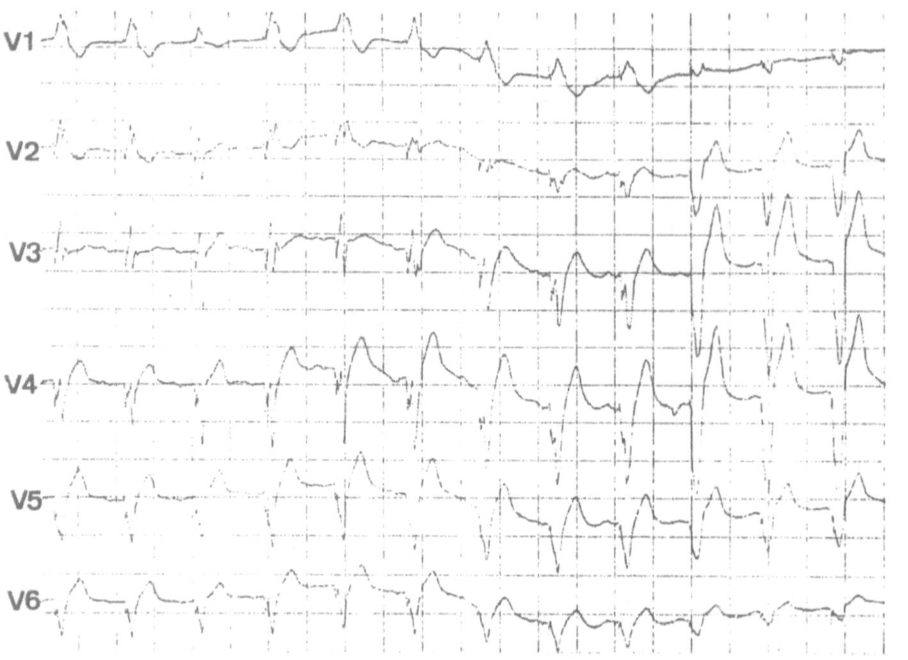

Figure 10. Ventricular epicardial pacing due to perforation.
Precordial leads from a patient with a unipolar lead perforated through the right ventricular apex. The third beat from the left is a fusion beat. The QRS complex on the left is positive in V1 and V2 due to the perforation. On pull-back, the polarity of the complex gradually returns to the classic negative pattern in V1 when the electrode tip resumes its position within the right ventricular apex.

the pacemaker-sensing circuitry. Although the pacemaker spike may, to some extent, deform the QRS complex, it occurs too late to cause true fusion, because it is delivered within the refractory period of the ventricle [28] (Figure 12).

The superimposition of an atrial stimulus upon the QRS complex may produce beats resembling pseudo-fusion beats because an atrial pacemaker spike is delivered after the inscription of the onset of the surface QRS but before the occurrence of the intrinsic deflection in the endocardial QRS. This phenomenon has been coined a pseudo-pseudo-fusion beat [29].

In dual chamber pacemakers, apparent undersensing of the QRS complex may occur after delivery of an atrial stimulus:

1. in DDD pacemakers if the ventricular complex occurs during the blanking period of the ventricular channel, which is started with each atrial stimulus in order to prevent crosstalk.

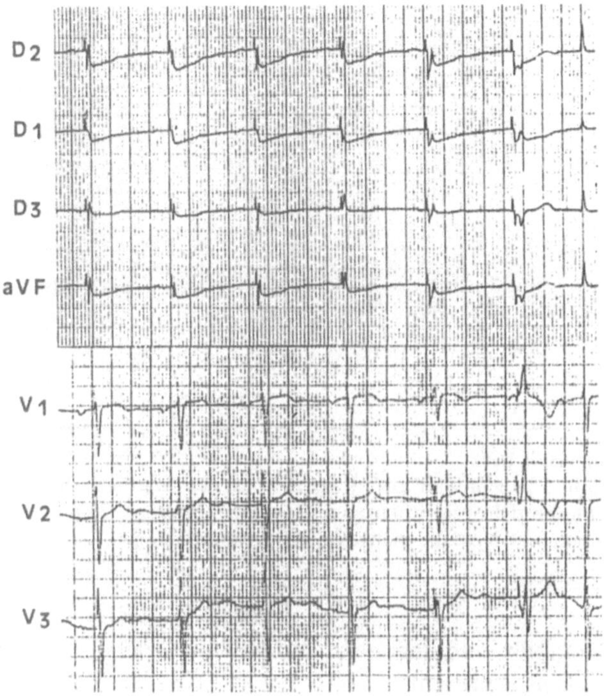

Figure 11. Apparent loss of capture.
Simultaneous recording of four telemetry leads (upper panel) and three precordial leads (lower panel) in a patient shortly after implantation of a VVI pacemaker. From the lower panel normal pacemaker function is evident with pseudofusion beats and one fusion beat (sixth beat from the left). Due to the automatic gain compensation built in the telemetry system, the amplitude of the native complexes is decreased thereby creating the impression of non-capture.

2. in committed DVI pacemakers which are refractory to ventricular events, after delivery of an atrial stimulus throughout the AV interval.

Some form of apparent undersensing may actually result from oversensing and may be corrected by decreasing the sensitivity of the pulse generator. When continuously exposed to electromagnetic interference or myopotentials, most pacemakers are protected against prolonged inhibition by automatic reversion to fixed-rate pacing [30], the so-called interference reversion mode. Decreasing the sensitivity may restore the demand function [31]. The interference or reversion mode may also be activated by atrial fibrillation with fast ventricular response [31].

2.2.3.3. *Oversensing.* If the interval between two stimuli or between a

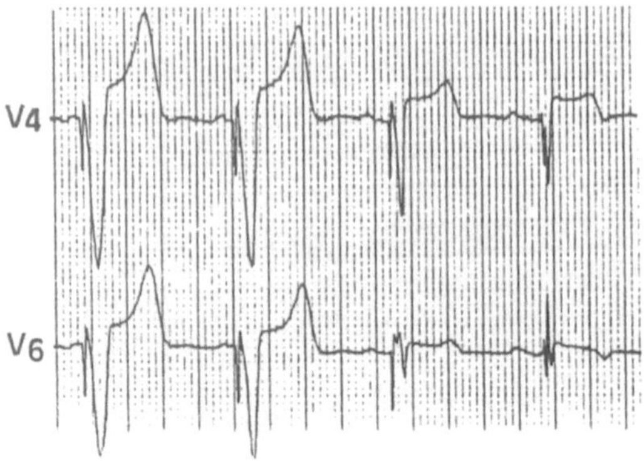

Figure 12. Pseudofusion beat.
Lead V4 and V6 in a patient with a VVI pacemaker. The first beat shows a normal paced QRS complex form the right ventricular apex. The third beat is a fusion beat and the fourth beat is a pseudofusion beat. Note the changes in T wave morphology.

spontaneous complex and the next stimulus is longer than the automatic interval, oversensing is present.

Oversensing can be easily diagnosed and eliminated by application of a magnet which converts the pacemaker to the asynchronous mode. The nature of the signal responsible for the oversensing can sometimes be elucidated by telemetry of the intracardiac electrogram or with the help of the marker channels.

2.2.3.4. *Pacemaker pauses.* Oversensing constitutes by far the most common cause of pacemaker pauses encountered clinically and may be due to:

1. *Signals set up in the pacing system independent of cardiac activity.* These signals result from:
   - *polarization* at the electrode-tissue interface or
   - *false signals* generated by a loose connection or wire fracture with intermittent contact. These signals are often enhanced by biologic signals such as P or T waves.

Delivery of the pacemaker pulse charges the electrode tissue interface to a large DC potential which is dissipated over a relatively long period. The resulting time-changing voltage represents an electric signal that may be sensed by the sensing circuit of the pacemaker when the refractory period is over. The phenomenon occurs more often with larger voltage and longer duration of the output pulse, especially in conjunction with short refractory period and high sensitivity settings of the pulse generator [32, 33]. The sensing of the polarization voltage has been called the 'double reset phenomenon' [34]. It should be suspected whenever the interval between two consecutive spikes lengthens to a value approximately equal to the sum of the refractory period and the automatic interval. Incorporation of a 'fast recharge' circuit in the output stage of most modern pacemakers may practically eliminate the sensing of afterpotentials [37].

False signals (electrical transients) are set up by abrupt changes in resistance within the pacing system and may inhibit the pacemaker in a random fashion causing chaotic pacemaker pauses. False signals can result from intermittent loose connections, wire fracture with otherwise well-apposed ends, short circuits, e.g. in bipolar electrodes or in pacemaker connectors, or inactive electrodes with bared terminals which should therefore always be insulated with biocompatible material [35, 36].

False signals can sometimes be recognized on the surface electrocardiogram, especially when taken with high fidelity physiologic recorders. They can more easily be recorded on the intracardiac electrocardiogram, either preoperatively or by telemetry.

The recognition of false signals as a cause of pacemaker pauses and its differentiation from benign causes like T wave sensing is of utmost importance, as it may be the first manifestation of a potentially catastrophic disruption of the electrode or the electrode-pulse generator contact [37].

2. *Cardiac signals* such as P or T waves, concealed ventricular extrasystoles.

Sensing of *P waves* by properly positioned ventricular electrodes at the right ventricular apex rarely occurs due to the small voltage of the P wave in the right ventricular electrogram.

In view of the generally higher sensitivity settings of atrial sensing circuits, *sensing of QRS complexes by atrial pacemakers* or atrial channels of dual chamber pacemakers is more likely to occur. Far-field QRS complex sensing by atrial leads can be elicited in 35% of patients with unipolar DDD pacemakers and should be eliminated by appropriately programming atrial sensitivity, refractory period, and atrioventricular delay [39]. Far-field QRS-complex sensing in atrial pacemakers can give rise to inappropriate pacemaker bradycardia if QRS sensing occurs outside the refractory period or to apparent undersensing of P waves if QRS sensing occurs during the relative refractory period of the AAI pacemaker which may then revert to the fixed rate or reversion mode [40].

In dual chamber pacemakers, far-field QRS sensing outside the refractory period by the atrial amplifier may trigger ventricular pacing and induce pacemaker-mediated tachycardia without the need for retrograde ventriculo-atrial conduction [40].

T wave sensing occurs more frequently after paced beats than after spontaneous complexes, due to summation of the T wave with the after-potential. Increase in the amplitude of the T wave by a current of injury such as occurs in the first weeks after implant or with myocardial infarction, may increase the likelihood of T wave sensing and often resolves spontaneously [41, 42] (Figure 13).

Creation of pacemaker pauses due to concealed ventricular premature beats confined to the His–Purkinje system and, thus, invisible on the surface electrocardiogram remains a theoretical possibility [43, 44].

3. *Myopotentials*, often visible on the surface electrocardiogram, are reported to inhibit unipolar demand pacemakers in a wide proportion of patients but only a minority of these are symptomatic [45] (Figure 14).

Recent modifications of the sensing circuit and the use of coated pacemakers can reduce the incidence of myopotential inhibition [46].

With single chamber units, the problem can usually be solved by reprogramming a lower sensitivity, reprogramming to the asynchronous or triggered mode, or reprogramming from the unipolar to the bipolar stimulation mode, if feasible. Bipolar pacemakers indeed appear relatively immune to myopotential interference. In general, asymptomatic myopotential interference has to be differentiated from other causes of pacemaker inhibition and can better be left alone aside from programming to a lower sensitivity or the bipolar mode, as reprogramming to the VOO or VVT mode may induce serious arrhythmias if an arrhythmia substrate is present.

Rarely, myopotential sensing by a VVI pacemaker may activate the reversion mode [47]. The reversion mode is a constant or asynchronous

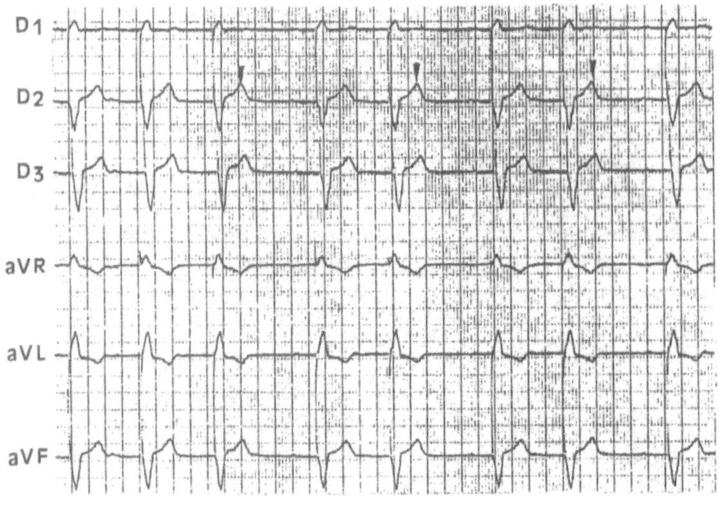

Figure 13. Oversensing due to T wave sensing.
Limb leads in a patient with a VVI pacemaker. Intermittent prolongation of the pacing interval occurs due to T wave sensing (arrows).

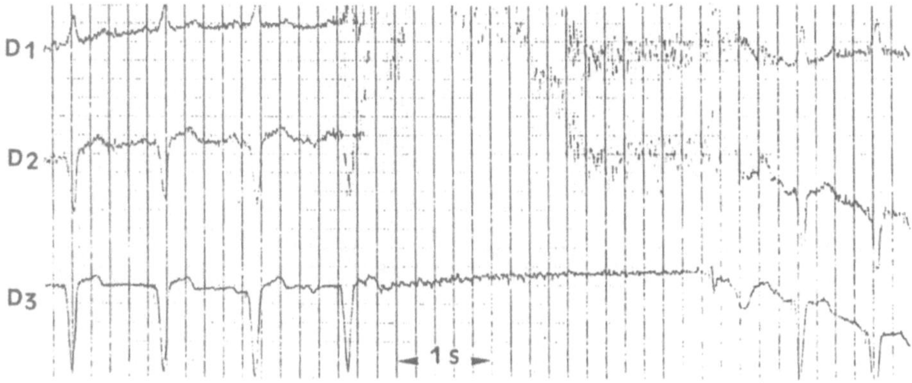

Figure 14. Myopotential inhibition of pacemaker output in a VVI unit.
Three standard ECG leads from a patient with a VVI pacemaker because of high degree atrioventricular block are shown. Pectoral muscle exercise inhibits pacemaker output with creation of a 3.9 s pause followed by a ventricular escape beat.

pacing mode intended to defend the pacemaker dependent patient against electromagnetic interference but it is, in general, ineffective against myopotential interference. With the advent of dual chamber pacemakers, myopotentials have been implicated not only in inhibition of pacemaker output but also in triggering ventricular output or induction of pacemaker-mediated tachycardia by simulating atrial activity [48] (Figure 15).

4. *Electromagnetic interference.* If a voltage across both terminals of a pacemaker is built up, reaching or exceeding a certain level, the timing circuit of the pacemaker is influenced and pacemaker pauses may be the result. If the interference signal is recognized as such, the pacemaker is switched to its interference mode or reversion mode which means asynchronous pacing as long as the state of interference persists [49].

The sources of interference are very diverse [49]; Irnich discerns five categories of exogenic interference:

a. *Galvanic interference.* Induced by direct current flowing through the body, e.g. touch activated switches, electrocautery, electric acupuncture, defective and not properly grounded domestic apparatus.
b. *Magnetically-coupled interference.* A time-varying magnetic field induces a voltage in the semicircular loop formed by the pacemaker and its electrode. Such magnetic fields are produced near electric welders, some types of electric razors, anti-theft devices, and weapon-detecting gates. However, the normal reaction to such gates is to inhibit the pacemaker

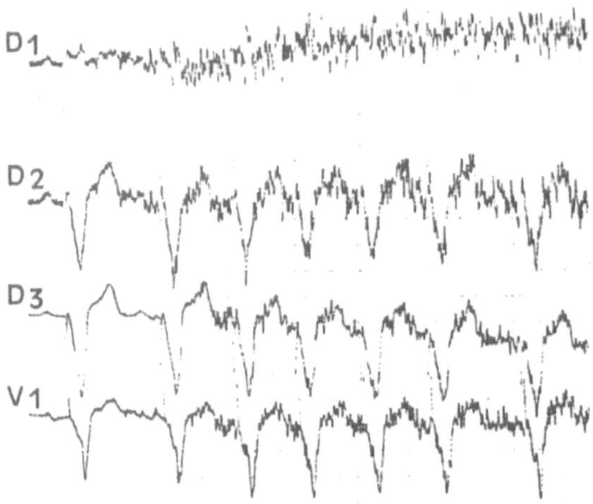

Figure 15. Myopotential triggering of ventricular pacing in a DDD unit.
Three standard ECG leads and lead V1 of a patient with a DDD pacemaker because of high degree atrioventricular block are shown. The first and second beat from the left show normal triggering of ventricular output by sinus P waves. From the third beat on, acceleration of ventricular pacing due to oversensing of myopotentials by the atrial channel occurs.

only for one beat when passing into the gate or to revert to the interference mode.

The strong magnetic field of magnetic resonance imaging generally closes the reed switch, thereby reverting the pacemaker to its asynchronous mode [50, 51]. However, in some models, rapid pacing was induced by the radiofrequency field of the scanners [50, 51].

c. *Electrically-coupled interference.* Each conductor or metallic body coupled to a grounded high-voltage source influences an electric charge on the surface of the human body comparable to a capacitor. If the distance is close enough and the voltage alternating with time, the pacemaker may be influenced. Examples are electric motors and ignition systems of internal combustion.
d. *Electromagnetically-coupled interference.* In the higher frequency range above 100 kHz, electric and magnetic fields cause each other and they can no longer be treated separately. Such waves can enter the body and influence the pacemaker, examples are diathermal treatment, microwave ovens, and strong radio transmitters.
e. *Magnetostatically-coupled interference.* Moving a strong magnet over the pacemaker can influence it. This is not due to an induced voltage but is the result of the activation of the reed switch. If the reed switch is incor-

porated into the sensing circuit, activation of the reed switch may induce a signal that inhibits the pacemaker [52].

5. Inhibition of pacemaker spikes by *other pacemaker spikes* as can occur with chest-wall stimulation an external pacemaker, and crosstalk in dual chamber devices [53, 54] (Figure 16).

6. Pulse *generator failure* with component failure occasionally associated with autointerference from internally sensed signals.

PM electrocardiogram in single chamber units:

1. no stimulus artefact can be identified
   * PM inhibition by EMI, native complexes, etc.
   * PM failure

   APPLY MAGNET
2. stimulus artefact can be identified

   observe:
   1. alterations and variability of spike amplitude
   2. alterations and variability of spike axis

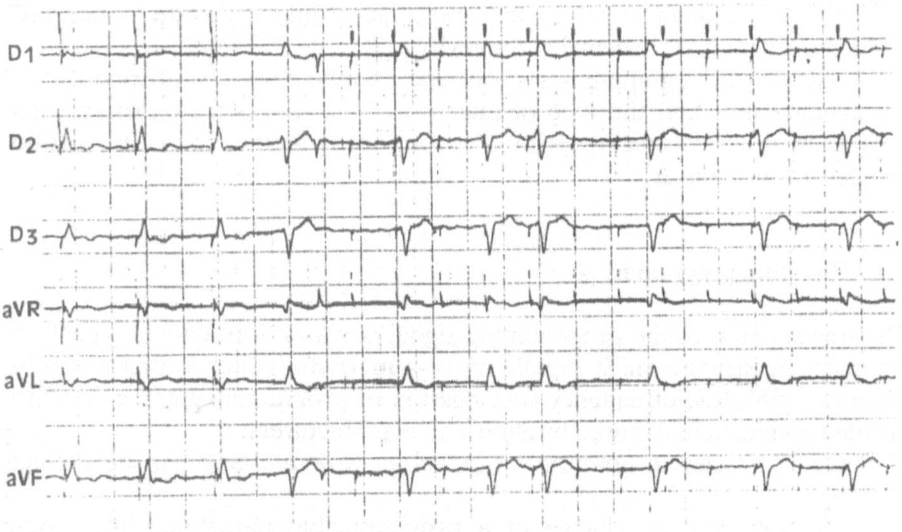

Figure 16. Inhibition of pacemaker activity by chest wall stimulation.
Limb leads in a patient with a VVI pacemaker. Note the abnormal frontal axis of the paced complex due to abnormal positioning of the electrode tip in the right ventricular outflow tract. The fourth beat is a native complex occurring after an interval slightly shorter than the automatic interval and thus inhibiting the pacemaker. Thereafter chest wall stimulation is initiated (arrows) that inhibits the pacemaker output temporarily.

3. is capture consistently present?
4. characteristics of the pacemaker evoked response
5. relation of the spike to the native complexes: under- and oversensing

3. MULTIPROGRAMMABILITY

Whereas multiprogrammability was considered a superfluous luxury at first by some, it has clearly become a very useful and, for DDD pacing, even an essential tool in order to provide optimal function of the pacing system.

Multiprogrammability offers many advantages [55]:

1. It allows optimal adjustment of the pacing system to each specific clinical situation. This may be especially useful in view of the increased lifetime of the present-generation pulse generators.
2. Multiprogrammable pulse generators have greatly simplified trouble-shooting of many pacemaker problems.
3. Many pacemaker complications can be treated by reprogramming thereby avoiding operative intervention. Surgical reintervention was necessary for pacemaker malfunction not related to lead dislodgment or perforation in 9% of nonprogrammable pulse-generator implants versus 3% of multiprogrammable implants [56].
4. Careful adjustment of pacemaker output according to threshold measurements can increase pacemaker longevity [57].

Before starting a programming session, one must be correctly informed about the manufacturer and the type of pacemaker that the patient carries, since programming a pacemaker with a wrong programmer may irreversibly damage the pacemaker.

3.1. Phantom programming

The transmission of the programming signal is via radiofrequency signals or via magnetic signals, which rapidly move a magnetic switch within the pulse generator [58]. Radiofrequency transmission of programming commands has become dominant and is used by almost all manufacturers.

Three categories of *spurious* or *phantom programming* can be defined [59, 60]

1. *dysprogramming*: change of a programmable parameter from some anomalous, nonpacemaker-related source or by inadvertent application of a noncorresponding programmer (*cross-programming*). Cold exposure during shipment (or during surgery with hypothermia?) may cause resetting of pulse generators. This phenomenon has been reported only for DDD pulse generators [61], but may probably also occur with single-chamber units. DDD units that are exposed to cold may be reset to the elective replacement indicator or back-up mode (either VVI or VOO mode according to the model and

manufacturer). Similar resettings to the back-up mode have been reported after electrocautery [62].

The units can simply be reprogrammed to the DDD mode if the problem is recognized as such. In some older units, however, the resetting of the pulse generator is not telemetered back to the programmer, creating a discrepancy between the actual behaviour of the pacemaker (VOO or VVI) and what is on the display of the programmer.

In our department, an attempt to mistakenly program a Pacesetter AFP 283 DDD pulse generator with a Cordis programmer resulted in the erasure of the identification code of the pulse generator which could not respond anymore telemetrically to the correct programmer interrogating the pulse generator. This resulted in reversion to the back-up VVI mode from which the pulse generator could not be resuscitated anymore to full DDD operation.

Similarly, a wrong identification code on a Cordis programmer (applying to another Cordis model) resulted in programming an inadvertent slow output of the pacemaker simulating end of life behaviour.

2. *misprogramming*: a generator response different from the intended response due to faulty program emission signals.

3. *phantom physician programming*: erroneous reprogramming due to human error, e.g. mistakes in the position of the switches of the programmer, nonregistered reprogrammings, etc.

It should be the aim of the manufacturers to make the pulse generators less vulnerable to extraneous energy sources by improving the selectivity of the transmission with insertion of an access code, by the use of binary language and by incorporating a final control of the message [59].

Fortuitously, dysprogrammed pacemakers can indeed hardly be differentiated from defective ones and, in fact, sometimes dysprogramming of the pacemaker can result in permanent damage to the pulse generator.

The most frequently programmed parameters are rate, output, and sensitivity. For dual-chamber pacemakers, extensive programmability is essential.

3.2. Output

Output is the most important programmable parameter and output programmability is the major reason for a decrease of the surgical reintervention rate after pacemaker implant [56]. Ideally, all pulse generators should be programmable in terms of both stimulus strength (voltage or current) and duration.

The energy content (W) of the impulse equals the product of current (I), voltage (U) and impulse duration (t) [63]:

$$W = U \cdot I \cdot t. \qquad (1)$$

It follows that the energy content of the impulse can be decreased either by diminishing the amplitude of current or voltage or by decreasing the impulse duration. However, energy can be conserved more efficiently by decreasing the voltage (or current) output of the pacemaker than by decreasing the pulse width.

Ohm's Law states

$$U = I \cdot R \text{ or } \quad I = U/R. \tag{2}$$

Substituting (2) into (1) yields

$$W = \frac{U^2 \cdot t}{R}. \tag{3}$$

From (3), it follows that the energy content of the pulse wave decreases linearly with pulse width but exponentially with pulse amplitude and, therefore, that energy conservation is most efficiently established by optimally decreasing the amplitude of the pulse, taking into account an adequate safety margin.

A *decrease of output* or energy content has several applications:

(1) Noninvasive determination of *pacing threshold* and safety margin. Nowadays, most pacemakers have an automated program to determine the threshold noninvasively by a stepwise decrease of either impulse height or duration.

For chronic lead implants, if the threshold is determined by decreasing the amplitude at constant pulse width, doubling the threshold value provides adequate safety. If, however, the threshold is determined by decreasing the impulse duration at constant amplitude, the pulse width must be adjusted to three times the threshold value for adequate safety [55].

(2) Increase of the *longevity* of the pulse generator taking into account threshold determination and an adequate safety margin. Energy conservation may be of particular importance in dual-chamber pacemakers and rate responsive systems.

(3) *Subthreshold stimulation* allows evaluation of the underlying rhythm, the spontaneous electrocardiogram (e.g. in the diagnosis of myocardial infarction), evaluation of *pacemaker dependency.* In some pacemakers, the same objective can be reached by temporary inhibition of pacemaker output or by temporary programming to the OOO mode.

(4) *Suppression of accessory stimulation*: often diaphragmatic or anodal muscle stimulation can be diminished by reducing the output of the pacemaker; reduction of impulse amplitude seems more effective than reduction of the duration [55] in order to avoid skeletal muscle stimulation.

An *increase of output* is mostly intended to overcome temporary or permanent increases in pacing threshold as may occur shortly after implantation (maturation effect), after DC shock or cardioversion, during myocardial infarction, electrolyte imbalance, or antiarrhythmic drug treatment.

3.3. Rate

A *decrease in the rate* of the pacemaker has several advantages:
(1) energy consumption of the pulse generator is decreased, thus increasing longevity;
(2) with VVI units, a decrease of the rate allows sinus rhythm to prevail, avoiding symptoms of the pacemaker syndrome if the patient is in sinus rhythm and has normal atrioventricular conduction most of the time. In some of these patients, programming to a low rate can even avoid the necessity of implanting a more complicated dual chamber system;
(3) in patients with coronary artery disease and angina pectoris, lowering the rate may decrease the oxygen demand of the heart;
(4) temporary lowering of the rate may unmask the underlying spontaneous rhythm and ECG, allowing the evaluation of pacemaker dependency and to establish a specific electrocardiographic diagnosis.

The issue of lowering the heart rate is thus mainly a problem of patients in sinus rhythm with intermittent bradycardia that are treated with a VVI pacemaker as AAI and dual-chamber pacemakers retain the normal atrioventricular sequence.

Therefore, for patients with only intermittent bradycardia, we adopt the following guidelines:
(1) if the patient has coronary artery disease and angina pectoris, the rate can be lowered to 40 or 50 bpm;
(2) if the patient has no angina pectoris, it is still advisable to lower the pacing rate to below the spontaneously prevailing rhythm, if the patient is in sinus rhythm. For patients in atrial fibrillation, the pacing rate can be left at the classic rate of 70 bpm.

Increasing the rate of the pacemaker to higher values than the nominal setting of 60 to 70 bpm is rarely necessary. The indications for rate increase are therefore limited:
(1) temporary rate increase can be indicated in order to augment the cardiac output (e.g. in case of fever, after surgery, etc). This may be useful in patients with an AAI or dual-chamber system or in patients in atrial fibrillation with a VVI system. In other patients, the benefit of increasing the rate must be weighted against the potentially deleterious effect of loosing normal atrioventricular sequencing;
(2) sometimes hemodynamically ineffective ventricular premature beats can be suppressed by programming to a higher rate.

3.4. Sensitivity

Sensitivity is a measure of the minimal potential difference required between the poles of the sensing amplifier in order to suppress or trigger the output of the pacemaker. Although sensitivity is expressed in millivolts, i.e. in terms of amplitude, the minimum requirements of adequate sensing are more complex and determined by the frequency content (rate of change of voltage of the signal or slew rate) of the signal as well. In the absence of a universally accepted and standardized test signal, sensitivity settings in various pacers cannot be directly compared.

Undersensing can be corrected by increasing the sensitivity of the pulse generator, i.e. reprogramming it to a lower numerical value whereas *oversensing* can be corrected by decreasing the sensitivity, i.e. reprogramming to a higher numerical value. Unsensed ventricular extrasystoles occur in 15% of patients with normal sensing of supraventricular complexes on routine Holter monitoring [31] and represent one of the limitations of the sensing circuit of contemporary pacemakers. It is therefore not indicated to correct this phenomenon by increasing the sensitivity of the pulse generator [31], unless the patient has evidence of an arrhythmia substrate (spontaneous or inducible ventricular tachycardia).

Undersensing is a frequent problem with atrial or dual-chamber pacemakers, as the amplitude of the atrial endocardial signal is generally smaller than the amplitude of the endocardial ventricular signal.

3.5. Refractory period

The refractory period constitutes the time interval following the delivery of an impulse (paced refractory period) or a sensed signal (sensed refractory period) during which the sensing amplifier is totally unresponsive to electrical signals.

The refractory period avoids inhibition of the pacemaker by its own impulse, the pacemaker evoked response, the T wave or polarization voltages generated at the electrode-tissue interface. Accordingly, if pacemaker pauses are generated by one of these factors, reprogramming of the refractory period to a longer duration may solve the problem.

In AAI units, the refractory period must be programmed to a relatively long value in order to prevent inhibition of pacemaker output by the ensuing QRS complex.

In DDD units, the atrial refractory period encompasses the atrioventricular delay and the post-ventricular atrial refractory period. The former avoids atrial evoked response sensing, the latter inhibition by the ventricular impulse or its evoked response in addition to an eventual retrograde atrial activation, which might otherwise induce pacemaker endless loop tachycardia.

Therefore, the main indications for *lengthening of the refractory period* include:

1. in *AAI* units in order to avoid inhibition by the ensuing QRS complex. The atrial refractory period should be close to 400 ms [64];
2. in *VVI* units in order to avoid inhibition by the T wave, by polarization artefact or both;

3. in *VDD* and *DDD* units in order to prevent endless loop tachycardia [65–67].

In most of the presently marketed models, additional lengthening of the postventricular atrial refractory period takes place after sensing a premature beat in one of the programmable features in order to prevent endless loop tachycardia [68, 69], which is often initiated by an extrasystole [70].

Shortening of the refractory period can be indicated when some beats are not sensed due to high rate (such as may occur with atrial fibrillation with fast ventricular response) or tightly coupled ventricular premature beats [64].

If a pacemaker is reprogrammed to the VVT or AAT mode in order to perform electrophysiologic studies in combination with chest wall stimulation, the refractory period should be appropriately shortened in order to allow high pacing rates to occur.

3.6. Polarity

Some pacemakers are programmable to either the unipolar or the bipolar mode [71, 72]. In more recent versions, sensing and pacing can be separately programmed to the unipolar or bipolar mode.

Reprogramming to *bipolar sensing* may be useful:

(1) to avoid myopotential inhibition and/or triggering;
(2) to avoid inhibition by electromagnetic interference.

Reprogramming to *unipolar sensing* is rarely indicated but may sometimes sufficiently enlarge a marginally small endocardial electrogram in order to establish normal sensing. On the rare occasion where one of the coaxial wires is broken, pacing and sensing can be restored by reprogramming to the unipolar mode.

Reprogramming to *bipolar pacing* can avoid pectoral muscle stimulation.

Reprogramming to *unipolar pacing* mode can restore capture in the above-mentioned situation of fracture or insulation failure of one of the coaxial wires. In this situation, the reprogramming to the unipolar mode is diagnostic. In addition, temporary programming to unipolar pacing can facilitate analysis of impulse wave form and measurement of impulse duration and interval.

3.7. Hysteresis

If the escape interval is different from the pace-to-pace interval, the pacemaker is considered to operate in the hysteresis mode. Most multiprogrammable pacemakers are nowadays provided with a positive hysteresis option, i.e. the escape interval is longer than the pace-to-pace interval, thus allowing the spontaneous rhythm to decrease below the programmed pacing rate. The positive hysteresis feature intends to take advantage of the spontaneous sinus rhythm, normal atrioventricular sequence, and ventricular activation. In practice, there is little benefit in comparison to simply programming a lower pacing rate without hysteresis [73]. In addition, the hysteresis complicates the electrocardiogram interpretation and several pitfalls have been reported.

3.8. Pacing mode

In single-chamber units (SSI) *conversion* from the inhibited *to the triggered mode* (SST), may be indicated in certain clinical circumstances:

(1) Oversensing due to myopotential inhibition, electromagnetic interference, etc. if symptomatic or in pacemaker dependent patients. Although programming to the SST mode does not actually correct the phenomenon of oversensing, the occurrence of long pauses can be prevented.
(2) Temporary programming to the synchronous mode can be useful, in combination with chest-wall stimulation, diagnostically (to perform electrophysiologic studies) and therapeutically (to interrupt various forms of tachycardia, e.g. ventricular tachycardia, supraventricular tachycardia, atrial flutter). For the interruption of atrial flutter, it is necessary to overdrive the flutter, i.e. the pacemaker must allow pacing at very high rates (above 300 bpm).

In dual chamber systems, programmability of pacing mode has become essential and is dealt with below.

3.9. Programmability in dual chamber systems

3.9.1. *Pacing mode*
On most models, a wide variety of pacing modes is available, some of which are essential.

VVI: can be an essential safety valve if permanent supraventricular tachycardias arise such as atrial fibrillation or flutter causing permanent rapid ventricular pacing in the DDD or VDD mode.

AAI: (or AOO) provides the most convenient way of testing for atrial capture in the presence of normal atrioventricular conduction.

VVT/AAT: are useful for electrophysiologic studies and in the treatment of tachyarrhythmias.

VOO/AOO: are similarly useful for termination of ventricular or supraventricular tachyarrhythmias respectively by under- or overdrive pacing.

DVI: as atrial sensing function is lacking, this mode constitutes one of the possible solutions for pacemaker mediated tachycardia, especially when the spontaneous atrial rate is low.

DVI (committed): can be indicated when cross-talk is present and atrial output cannot be lowered because of high atrial threshold levels.

VDD: can be programmed to avoid cross-talk or to avoid stimulation of phrenic nerve or pectoral muscle, if sinus function is relatively normal. Phrenic nerve stimulation may result from a malpositioned atrial J lead whereas pectoral muscle stimulation commonly results from an insulation break near the pulse generator, defective coating, connector problems or from the superfast atrial recharge pulse [74].

DAD: can rarely be a solution for inhibition of the ventricular impulse in pacemaker dependent patients due to false signals, or electromagnetic interference (Figure 17).

DDI: is mainly indicated in patients with paroxysmal atrial fibrillation or flutter, that are in sinus rhythm most of the time. It has been stated that the DDI mode is the mode of choice in patients with the sick sinus syndrome [75].

In the DDI mode, there is no synchronization of ventricular output to the sensed atrial signal avoiding rapid ventricular pacing with fast supraventricular rhythms. However, there occurs no increase of the paced ventricular rhythm on exercise and the ventricular stimulus can occur widely separated in time from the atrial depolarization, thus favouring the occurrence of retrograde ventriculo-atrial conduction, just as with VVI pacing.

The DDI mode has conceptually been considered as a DDD mode with an upper rate identical to the lower rate [76].

DDD pacing with upper rate smoothing may be another option in patients with the brady-tachy syndrome [77].

3.9.2. *Atrioventricular interval*

Most DDD pacemakers have identical atrioventricular intervals after sensing or delivery of an atrial stimulus resulting in sometimes widely different effective PR intervals. In some units, the interval after atrial pacing is appropriately lengthened. The optimum atrioventricular interval can most conveniently be determined with cardiac doppler methods [78] and may vary widely between patients [79]. An AV interval between 150 and 200 ms is most likely to provide the highest cardiac output.

For physically active patients, progressive shortening of the AV interval with increasing atrial rate may be of benefit.

Shortening of the AV interval in the pacemaker clinic ascertains constant ventricular depolarization by a ventricular stimulus and is of help to test for P wave tracking and cross-talk.

Lengthening of the AV interval in the presence of a normal AV conduction effectively produces pacing in the AAI mode thus conserving energy without loosing the safety of ventricular pacing standby. In addition, asynchrony of left ventricular contraction due to abnormal (PM induced) depolarization is avoided.

3.9.3. *Atrial refractory period*

The total atrial refractory interval (TARI) consists of the AV interval (AVI) and the postventricular atrial refractory period (PVARP) [80]. Often the programmer displays only the latter part of the atrial refractory period. *Lengthening* the PVARP to beyond the ventriculoatrial conduction time is one of the modalities to prevent pacemaker mediated endless loop tachycardia but will, at the same time, decrease the upper rate limit. The VA conduction time can be determined from the surface electrocardiogram, the telemetered endocardial electrocardiogram or with the help of telemetered marker channels.

The relation between the TARI and the upper rate interval (URI) set by the upper rate limit determines the upper rate behaviour [24]. If the TARI is equal to the URI the pacemaker will track the P waves at progressively shorter spontaneous atrial intervals (SAI) during exercise until the ventricular paced rate suddenly drops due to 2:1 block. If the TARI is shorter than the URI, the upper rate behaviour will be pseudo-Wenckebach, with a plateau at the upper rate limit before 2:1 block occurs.

3.9.4. *Upper rate limit*

The upper rate interval (URI) is the shortest possible interval between two consecutive ventricular stimuli. Similarly, the upper rate limit (URL) is the highest atrial rate that can be tracked in a 1:1 relation:

$$\text{URL (ppm)} = 60000/\text{URI (ms)},$$

$$\text{URI (ms)} = 60000/\text{URL (ppm)}.$$

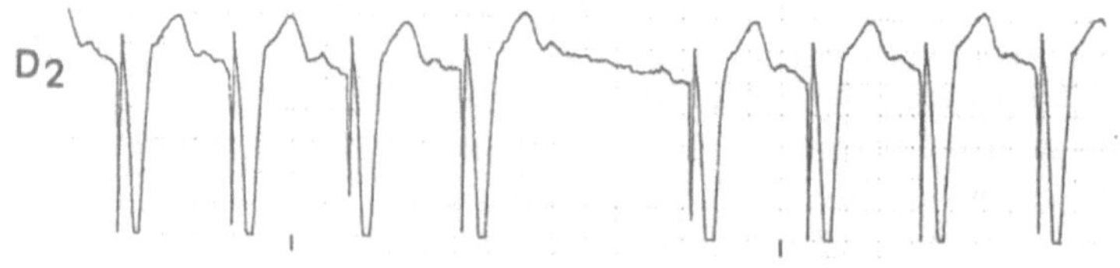

Figure 17. Ventricular oversensing causing pacemaker inhibition.
This 85 year old patient received a Cordis Sequicor 233F DDD unit in 1983 because of intermittent total atrioventricular block with recurrent syncope. Although after the implant on every occasion he proved to be completely pacemaker-dependent, he remained well until March 1985, when he complained of recurrent dizziness without syncope. On extensive pacemaker evaluation, the unit showed entirely normal function. On Holter monitoring, recurrent episodes were seen with missing ventricular stimuli, apparently due to false inhibition of ventricular output. However, we could not identify myopotentials or some other form of electromagnetic interference as a cause of the inhibition. The phenomenon and the symptoms disappeared after reprogramming the unit from the DDD to the DAD mode.

Limitation of the upper rate is indicated in patients with ischemic heart disease and with paroxysmal supraventricular tachycardias.

3.9.5. *Lower rate limit*
The lower rate limit (LRL) determines the lower rate interval (LRI), and consists of the atrioventricular interval (AVI) and the ventriculo-atrial interval (VAI). If the spontaneous atrial interval lengthens to a value beyond the LRI, a DDD, DAD, and DDI pacemaker start pacing the atrium and the ventricle.

In a VDD pacemaker, however, only the ventricle is paced, thereby inducing atrioventricular dissociation and increasing the likelihood of retrograde ventriculoatrial conduction and pacemaker mediated tachycardia.

3.9.6. *Blanking interval*
The blanking interval is designed to avoid sensing pacemaker stimuli, directed at one chamber, in the opposite chamber, whereas the refractory interval is designed to avoid sensing physiologic electric activity [81].

Blanking of the atrial amplifier occurs with ventricular pacing and is not programmable, as the ventricular stimulus always starts a postventricular atrial refractory period.

Blanking of the ventricular amplifier occurs with atrial stimulation and should be ideally programmable as during the ensuing atrioventricular interval the ventricular amplifier is not refractory. It should be programmed to the shortest possible value without provoking cross-talk, i.e. inhibition of ventricular output by the atrial stimulus.

4. INTERACTION WITH DIAGNOSTIC AND THERAPEUTIC DEVICES

Some diagnostic or therapeutic devices can profoundly influence pacemaker function or even damage the pacemaker. It is the purpose of this brief review to summarize the interaction of various diagnostic and therapeutic devices with pacemakers.

4.1. Cardioversion, defibrillation, electrocautery

Various adverse effects of cardioversion, defibrillation or electrocautery are known to occur:

a. The high current flow through the pulse generator may irreversibly *damage* [82, 83] the delicate electronic circuitry, although in modern pacemakers the circuitry is protected to some extent [84].

b. High current flows through the electrode-biointerface may cause electrocautery *burns* in the myocardium resulting in transient or permanent *rises in stimulation threshold* and/or *sensing disturbances*. For high current flow through the electrode to occur direct contact of electrocautery or defibrillation paddles with the pacemaker or lead is not required as capacitative coupling can induce major current flow through pacemaker leads [85].

c. *Reprogramming* of VVI units has been reported following DC shock [86], as well as *resetting to the back up mode* in DDD units [87].

d. Electrocautery has been reported to induce ventricular fibrillation in pacemaker patients [88].

Whenever possible, cardioversion or electrocautery should therefore be avoided in pacemaker patients.

Certain recommendations have been formulated for cardioversion and defibrillation in pacemaker patients [83]:

(1) the paddles should be kept as far away as possible from the pulse generator;
(2) if possible anterior-posterior paddle orientation should be used;
(3) the lowest possible energy should be used and energy must be titrated upto the accomplishment of the desired effect;
(4) after the DC shock pacemaker function and threshold behaviour should be carefully reevaluated;
(5) if the patient is pacemaker dependent it is prudent to have a temporary pacemaker and venous access route standby.

Similarly, after major surgery with use of electrocautery, it is prudent to reevaluate pacemaker function carefully.

Recommendations

(1) Keep paddles away from pulse generator, anterior-posterior location is preferable.
(2) Monitor EKG during electrocautery, eventually apply magnet.
(3) Carefully reevaluate pacemaker function after DC shock.

4.2. Ionizing radiation

The problem with ionizing radiation equipment is two fold [89].

a. *Electromagnetic interference*, mainly with the linear accelerator may theoretically influence pacemaker behaviour. Electromagnetic interference does not constitute a problem with the older cobalt irradiation equipment. Although only brief inhibition of pacemaker output due to electromagnetic interference resulting from on and off switching the linear accelerator has been reported [90], it is advisable to protect the pacemaker dependent patient subjected to treatment with a linear accelerator by continuous monitoring, eventually by reprogramming to a lower sensitivity or reversion to fixed rate by application of a magnet.

b. *Damage to the circuitry* may result from direct destruction of the semiconductor circuitry. The effect of ionizing radiation is cumulative and unpredictable. Sudden complete failure occurs in a substantial number of pulse generators subjected in vitro to the effects of a linear accelerator in doses that are comparable to those encountered clinically [91]. Transient recovery of function may be followed by total failure, suggesting that even a transient loss of function must be regarded as a precursor of permanent damage. [92].

Therefore, direct irradiation of the pulse generator must be avoided. If the pacemaker is in or near the radiation field, it must be shielded by a lead block, and EKG monitoring during the procedure is advisable in order to detect transient or permanent loss of output. Eventually the pacemaker must be replaced or relocated into another area.

Recommendations

(1) With a linear accelerator (not with telecobalt) electromagnetic interference may inhibit pacemaker output. Eventually reprogram a lower sensitivity or apply a magnet.

(2) If the pacemaker is within the radiation field it must be protected by a lead block or eventually relocated into another area.

(3) During the procedure EKG monitoring is advised to detect electromagnetic interference or transient or permanent loss of output due to radiation damage.

4.3. Nuclear magnetic resonance imaging (MRI)

In general, exposing patients with implanted metallic devices to MRI scanners is strictly contraindicated.

MRI scanners can influence cardiac pacemakers in several ways.

(1) In MRI scanners a very strong magnetic field is operative. The strength of the magnet is typically 0.35—0.50 Tesla, an enormous strength in comparison to the 0.0004 Tesla of the typical magnet that closes the reed switch of contemporary pacemakers [93]. All implantable pulse generators reverted to the asynchronous mode within the magnetic field of a conventional MRI scanner [94], when tested in vitro. This was confirmed later on in an in vivo dog mode [95]. Asynchronous pacing could be prevented, however, when the pacemaker was programmed to the 'magnet off' mode, which is not always possible [95].

(2) Theoretically, the very strong magnetic field could exert catastrophic attractive forces on metallic objects such as pulse generators and leads, but preliminary data indicate that the pull and torque exerted on the pacemaker in a dog model were inconsequential, even in the center of the magnetic field [94].

(3) The MRI scanner contains a radiofrequency source which may or may not be gated to the cardiac cycle. Rapid pacing even at rates above the rate protection limit of the pacemaker, both in VVI and in DDD units has been reported in experimental circumstances [95].

(4) The radiofrequency source can induce electromotive forces and current

flow within the pacemaker and lead, causing heating and potentially damage of the components.

Recommendations
Exposure of pacemaker patients to MRI scanner is contraindicated. If it is unavoidable reprogram to the magnet off mode.

4.4. Ultrasound

The primary mechanism for biological damage from ultrasound is from bulk heating. Because the energy generated in diagnostic ultrasound is limited to low values no appreciable tissue or pacemaker heating occurs [93].

4.5. Diathermy

Diathermy hyperthermia transmits electromagnetic waves of 50—500 MHz into the body to produce heating [93]. Diathermy should therefore be avoided as it can cause excessive heating of the pacemaker and can derange the demand function of the pacemaker by electromagnetic interference.

Recommendation
Diathermy is contraindicated in pacemaker patients.

4.6. Extracorporeal shock wave lithotripsy

The lithotriptor employs an underwater spark gap to generate a shock wave that is focused by an ellipsoidal metal reflector. Serial shock waves are applied until the stone is pulverized. Sometimes upto 1500 shocks are needed. The induction of ventricular extrasystoles can be avoided by synchronization of the shocks to the QRS signal [96].

The presence of a cardiac pacemaker has been considered a contraindication for treatment with the lithotriptor due to possible electromagnetic interference from the spark gap or damage to the pulse generator from the shock waves themselves.

Fifty percent of pacemaker systems exposed to shock waves in vitro were indeed inhibited when subjected to shock waves at a rate above the escape rate [97]. When, however, shock waves were delivered synchronously with pulse generator output, no inhibition was seen, and only one out of 22 pulse generators intermittently reverted to the magnet rate.

Another study comparing VVI and DDD units in vitro did not demonstrate any problem with VVI units when shock waves were synchronized to the ventricular stimulus. In DDD units, however, the shock waves synchronized to the atrial stimulus, which could inhibit the ventricular stimulus in some units [98].

This problem could be solved by decreasing the ventricular sensitivity or by temporarily reprogramming to the VVI mode.

The same investigators also demonstrated destruction of the piezoelectric crystal of rate responsive systems when the pacemaker was placed in the focal point of the lithotriptor. When the units were placed a few centimeters further away, they started pacing at their upper rate limit during shock wave lithotripsy due to activation of the piezo-electric element. Therefore, it can be concluded that extracorporeal shock wave lithotripsy is in general safe for pacemaker patients when certain precautions are taken into account. It is advised to reprogram DDD units to the VVI mode, or at least to a low ventricular sensitivity level. Patients with a piezoelectric activity sensing element should this feature have programmed off during shock wave lithotripsy and if such a pulse generator is implanted in the abdominal region is probably safer not to perform extracorporeal shock wave lithotripsy.

Recommendations

(1) Synchronize shock waves to pacemaker output.
(2) Reprogram DDD units to the VVI mode.
(3) Program rate responsive feature 'off', if activity sensing is based on a piezoelectric crystal.
(4) Do not perform extracorporeal shock wave lithotripsy if an activity sensing pacemaker with piezoelectric crystal is implanted in the abdominal region.

REFERENCES

1. Furman S (1985) Pacemaker follow-up, pp. 889—918 in Barold SS (ed.), *Modern Cardiac Pacing.*
2. Barold SS, Mugica J, Falkoff MD, Ong LS, Heinle RA (1985) Multiprogrammability in cardiac pacing, pp. 377—409 in Barold SS (ed.), *Modern Cardiac Pacing.*
3. Dressler L, Gruse G, von Knorre GH, Otte KB, Podszuz G, Richwien R, Schaedel H, Weber D, Weber D, Witte J (1979) The optimization of the pulse delivered by the pacemaker. *PACE* 2: 282—288.
4. Stokes K, Bornzin G (1985) The electrode-biointerface: stimulation, pp. 33—77 in Barold SS (ed.), *Modern Cardiac Pacing.*
5. den Dulk K, Lindemans F, Wellens HJJ (1984) Management of pacemaker circus movement tachycardias. *PACE* 7: 346—355.
6. Brandt J, Attewell R, Fahraeus T, Schuller H (1990) Acute atrial endocardial P wave amplitude and chronic sensitivity requirements: relation to patient age and presence of sinus node disease. *PACE* 13: 417—424.
7. Kruse I, Ryden L, Ydse B (1979) Clinical and electrophysiological characteristics of a transvenous atrial lead. *Br Heart J* 42: 595—602.
8. Sykosch HJ, Pletschen B, Thornander H, *et al.* (1987) Post-implant evaluation of detected P-wave amplitude (abstract). *PACE* 10: 730.
9. Zimmern SH, Clark MF, Duncan JL, *et al.* (1988) Variation of transmitted atrial electro-

grams following permanent atrial lead implantation. An intensive follow-up study (abstract). *PACE* 11: 533.
10. Van den Berg JW, Rodrigo FA, Thalen HJ, Koops J (1967) Photo-analysis of the condition of implanted pacemakers and electrode circuits. I. *Proc. Koninkl. Nederl. Akademie v. Wetenschappen, series C.* 70: 419—433.
11. Van den Berg JW, Rodrigo FA, Thalen HJ, Koops J (1967) Photo-analysis of the condition of implanted pacemakers and electrode circuits. II. *Proc. Koninkl. Nederl. Akademie v. Wetenschappen, series C.* 70: 434—447.
12. Furman S (1985) Pacemaker follow-up, pp 889—919 in Barold SS (ed.), *Modern Cardiac Pacing.*
13. Sinnaeve A, Willems R, Stroobandt R (1982) Inhibition of on demand pacemakers by magnet waving. *PACE* 5: 878— 890.
14. Siemens Pacesetter User Manual, model AFP 283.
15. Barold SS, Falkoff MD, Ong LS, Heinle RA (1985) The abnormal pacemaker stimulus, pp. 571—586 in Barold SS (ed.), *Modern Cardiac Pacing.*
16. Engler RL, Goldberger AL, Bhargava V, Kapelusznik D (1982) Pacemaker spike alternans: An artifact of digital signal processing. *PACE* 5: 748—750.
17. Barold SS, Falkoff MD, Ong LS, Heinle RA (1981) Electrocardiographic diagnosis of pacemaker malfunction, pp. 236—299 in Wellens HJJ and Kulbertus HE (eds), *What's New in Electrocardiography?*
18. Sinnaeve A, Willems R, Backers J, Holvoet G, Stroobandt R, (1987) Pacing and sensing: How can one electrode fulfill both requirements? *PACE* 10: 546—554.
19. Van Mechelen R, Vandekerckhove Y (1986) Atrial capture and dual chamber pacing. *PACE* 9: 21—25.
20. Nanda NC and Barold SS (1982) Usefulness of echocardiography in cardiac pacing. *PACE* 5: 222—237.
21. Nanda NC, Bhandari A, Barold SS, Falkoff M (1983) Doppler echocardiographic studies in sequential atrioventricular pacing. *PACE* 6: 811—814.
22. Hardebeck CJ (1988) Electrocardiographic characteristics of pacing from the right atrial appendage during atrioventricular sequential pacing. *PACE* 11: 193—202.
23. Castellanos A Jr, Ortiz JM, Pastis N, Castillo C (1970) The electrocardiogram in patients with pacemakers. *Prog Cardiovasc Dis* 13: 190—209.
24. Barold SS, Falkoff MD, Ong LS, Heinle RA (1985) The abnormal pacemaker stimulus, pp. 571—586 in Barold SS (ed.), *Modern Cardiac Pacing.*
25. Luceri RM, Furman S, Hurzeler P, Escher DJW, (1977) Threshold behaviour of electrodes in long term ventricular pacing. *Am J Card* 40: 184—188.
26. Furman S, Hurzeler P, De Caprio V (1977) Cardiac pacing and pacemakers III. Sensing the cardiac electrogram. *Am Heart J* 93: 794—801.
27. Driller J, Barold SS, Parsonnet V (1976) Normal and abnormal function of the pacemaker magnetic reed switch. *J Electrocardiol* 9: 283—292.
28. Spritzer RC, *et al.* (1969) Arrhythmias induced by pacemaking on demand. *Am Heart J* 77: 619—627.
29. Barold SS, Falkoff MD, Ong LS, Heinle RA (1981) Interpretation of electrocardiograms produced by a new unipolar multiprogrammable 'committed' AV sequential demand (DVI) pulse generator. *PACE* 4: 692—708.
30. Falkoff M, Ong LS, Heinle RA, Barold SS (1978) The noise sampling period: A new cause of apparent sensing malfunction of demand pacemakers. *PACE* 1: 250—253.
31. Van Gelder LM, El Gamal MIH (1988) Undersensing in VVI-pacemakers detected by Holter monitoring. *PACE* 11: 1507—1511.
32. Barold SS, Ong LS, Heinle RA (1980) Demand pacemakers: Normal and abnormal mechanisms of sensing , pp. 551—601 in Samet P, El-Sherif N (eds), *Cardiac Pacing.*
33. Hauser GH, Susmano A (1981) Afterpotential oversensing by a programmable pulse generator. *PACE* 4: 391—395.

34. Barold SS, Carroll M (1972) Double reset of demand pacemakers. *Am Heart J* 84: 276—277.
35. Yokoyama M, Hori M, Grechko M (1978) Suppression of demand mechanism by inactive myocardial electrodes. *PACE* 1: 126—131.
36. Chandra MS, Patel MR, Laughlin DE, Rossi NP (1978) 'False inhibition' of demand pacemaker due to leakage of fluid into the pacemaker lead socket. *J Thorac Cardiovasc Surg* 75: 765—768.
37. Barold SS, Falkoff MD, Ong LS, Heinle RA (1985) Differential diagnosis of pacemaker pauses, pp. 587—614 in Barold SS (ed.), *Modern Cardiac Pacing.*
38. Van Gelder LM, El Gamal MIH, Tielen CHJ (1988) P wave sensing in VVI pacemakers: Useful or a problem? *PACE* 11: 1413—1418.
39. Brandt J, Fahraeus T, Schuller H. (1988) Far-field QRS complex sensing via the atrial pacemaker lead. II. Prevalence, clinical significance and possibility of intraoperative prediction in DDD pacing. *PACE* 11: 1540—1544.
40. Brandt J, Fahraeus T, Schuller H (1988) Far-field QRS complex sensing via the atrial pacemaker lead. I. Mechanism, consequences, differential diagnosis and countermeasures in AAI and VDD/DDD pacing. *PACE* 11: 1432—1438.
41. Berman ND, (1980) T wave sensing with a programmable pacemaker. *PACE* 3: 656—659.
42. Yokoyama M, Wada J, Barold SS (1981) Transient early T wave sensing by implanted programmable demand pulse generator. *PACE* 4: 68—74.
43. Massumi RA, Mason DT, Amsterdam EA, Salel AF (1972) Apparent malfunction of demand pacemaker caused by nonpropagated (concealed) ventricular extrasystoles. *Chest* 61: 426—431.
44. Levine PA, Pirzada FA (1981) Pacemaker oversensing: A possible example of concealed ventricular extrasystoles. *PACE* 4: 199—203.
45. Breivik K, Ohm O (1980) Myopotential inhibition of unipolar QRS-inhibited (VVI) pacemakers, assessed by ambulatory Holter monitoring of the electrocardiogram. *PACE* 3: 470—478.
46. Fetter J, Bobeldyk GL, Engelman FL (1984) The clinical incidence and significance of myopotential sensing with unipolar pacemakers. *PACE* 7: 871—881.
47. Erkkila KI, Singh JB (1985) Reversion mode activation by myopotential sensing in a ventricular inhibited demand pacemaker. *PACE* 8: 50—51.
48. Den Dulk K, Lindemans F, Wellens HJJ (1984) Management of pacemaker circus movement tachycardias. *PACE* 7: 346—355.
49. Irnich W (1984) Interference in pacemakers. *PACE* 7: 1021—1048.
50. Holmes DR, Hayes DL, Gray JE, Merideth J (1986) The effects of magnetic resonance imaging on implantable pulse generators. *PACE* 9: 360—370.
51. Fetter J, Aram G, Holmes DR, Gray JE, Hayes DL (1984) The effects of nuclear magnetic resonance imagers on external and implantable pulse generators. *PACE* 7: 720—727.
52. Barold S, Gaidula J, Castillo R (1973) Unusual response of demand pacemakers to magnets. *Br Heart J* 35: 353—358.
53. Levine PA, Venditti FJ, Podrid PJ, Klein MD (1988) Therapeutic and diagnostic benefits of intentional crosstalk mediated ventricular output inhibition. *PACE* 11: 1194—1201.
54. Beaver BB, Maloney JD, Castle LW, Morant VA, Keefe JM, and Ching E (1986) Design-dependent crosstalk in a second generation DDD pacemaker. *PACE* 9: 65—77.
55. Barold SS, Mugica J, Falkoff MD, Ong LS, Heinle RA (1985) Multiprogrammability in cardiac pacing, pp. 377—409 in Barold SS (ed.), *Modern Cardiac Pacing.*
56. Pless R, Simonsen E, Arnsbo P, Fabricius J (1986) Superiority of multiprogrammable to nonprogrammable VVI pacing: a comparative study with special reference to management of pacing system malfunction. *PACE* 9: 739—744.
57. Furman S (1989) Pacemaker longevity. *PACE* 12: 1437—1438.

58. Hayes DL (1986) Programmability, pp.219—251 in Furman S. et al, (eds). *A Practice of Cardiac Pacing.*
59. Sinnaeve A, Piret J, Stroobandt R (1980) Potential causes of spurious programming. *PACE* 3: 541—547.
60. Cameron JR, Chisholm AW, Froggatt GM, Harrison AW (1979) Phantom Programming, pp. 35—3 in *6th World Symposium on Cardiac Pacing, Proceedings, PACESYMP.*
61. Barold SS, Falkoff MD, Ong LS, Heinle RA, Willis JE (1988) Resetting of DDD pulse generators due to cold exposure. *PACE* 11: 736—743.
62. Belott PH, Sands S, Warren J (1984) Resetting of DDD pacemakers due to EMI. *PACE* 7: 169—172.
63. Stroobandt R, Willems R, Depuydt P, Sinnaeve A (1982) Multiprogrammeerbare pacemakers: klinische aspecten. *Tijdschrift v. Geneesk.* 38: 597—604.
64. Barold SS, Ong LS, Heinle RA (1980) Demand pacemakers: normal and abnormal mechanisms of sensing, pp. 551—601 in Samet P, El-Sherif N (eds), *Cardiac Pacing,* 2nd edn, Grune and Stratton.
65. Furman S, Fisher JD (1982) Endless loop tachycardia in an AV universal (DDD) pacemaker. *PACE* 5: 486—489.
66. Luceri RM, Castellanos A, Zaman L, Myerburg RJ (1983) The arrhythmias of dual chamber cardiac pacemakers and their management. *Ann Intern Med* 99: 354—359.
67. Den Dulk K, Lindemans FW, Wellens HJJ (1986) Merits of various antipacemaker circus movement tachycardia features. *PACE* 9: 1055—1062.
68. Duncan JL, Clark MF (1988) Prevention and termination of pacemaker-mediated tachycardia in a new DDD pacing system (Siemens-Pacesetter model 2010T). *PACE* 11: 1679—1683.
69. Levine PA (1983) Postventricular atrial refractory periods and pacemaker mediated tachycardias. *Clin Prog Pacing Electrophysiol* 1: 394—401.
70. Den Dulk K, Lindemans FW, Bar FW, Wellens HJJ (1982) Pacemaker related tachycardias. *PACE* 5: 476—485.
71. Breivik K, Ohm OJ, Engedal H (1983) Long-term comparison of unipolar and bipolar pacing and sensing using a new multiprogrammable pacemaker system. *PACE* 6: 592—600.
72. Smyth NPD, Sager D (1983) A multiprogrammable pacemaker with unipolar or bipolar option. *Am Heart J* 106: 412—414.
73. Rosenqvist M, Vallin HO, Edhag KO (1984) Rate hysteresis pacing: how valuable is it? A comparison of the stimulation rates of 70 and 50 beats per minute and rate hysteresis in patients with sinus node disease. *PACE* 7: 332—340.
74. Stroobandt R, Willems R, Depuydt P, Holvoet G, Sinnaeve A (1989) The superfast atrial recharge pulse: a cause of pectoral muscle stimulation in patients equipped with a unipolar DDD pacemaker. *PACE* 12: 451—455.
75. Markewitz A, Schad N, Hemmer W, Bernheim C, Ciavolella M, Weinhold C (1986) What is the most appropriate stimulation mode in patients with sinus node dysfunction? *PACE* 9: 1115—1120.
76. Barold SS (1987) The DDI mode of cardiac pacing. *PACE* 10: 480—484.
77. Smisson DC (1986) Use of DDD rate smoothing to control sinus brady/tachy: a case study. *Clin Progr Electrophysiol Pacing* 4: 153—157.
78. Haskell RJ, French WJ (1988) Optimum AV interval in dual chamber pacemakers. *PACE* 9: 670—675.
79. Forfang K, Otterstad JE, Ihlen H (1986) Optimal atrioventricular delay in physiological pacing determined by doppler echocardiography. *PACE* 9: 17—20.
80. Stroobandt R, Willems R, Holvoet G, Backers J, Sinnaeve A (1986) Prediction of Wenckebach behavior and block response in DDD pacemakers. *PACE* 9: 1040—1046.
81. Furman S (1985) Dual chamber pacemakers: Upper rate behaviour. *PACE* 8: 197—214.
82. Aylwards P, Blood R, Tonkin A (1979) Complications of defibrillation with permanent pacemaker in situ. *PACE* 2: 462—000.

83. Levine PA, Barold SS, Fletcher RD, Talbot P (1983) Adverse acute and chronic effects of electrical defibrillation and cardioversion on implanted unipolar cardiac pacing systems. *J Am Coll Cardiol* 1: 1413—1422.
84. Lau FYK, Bilitch M, Weintroub HJ (1969) Protection of implanted pacemakers from excessive electrical energy of DC shock. *Am J Cardiol* 23: 244—249.
85. Shepard RV, Russo AG, Breland VC (1979) Radiofrequency electrocoagulator hemostasis in chronically elevated pacing thresholds in cardiopulmonary bypass procedure patients. In Meere C (ed.), *Cardiac Pacing: Proceedings of the 6th World Symposium on Cardiac Pacing*, Montreal Pacesymp. 35—2.
86. Barold SS, Ong LS, Scovil J, et al. (1978) Reprogramming of implanted pacemakers following external defibrillation. *PACE* 1: 514—520.
87. Belott PH, Sands S, Warren J (1984) Resetting of DDD pacemakers due to EMI. *PACE* 7: 169—172.
88. Hungerbuhler RF, Swope JP, Reves TG (1974) Ventricular fibrillation associated with use of electrocautery. *JAMA* 230: 432—435.
89. Venselaar JLM, Van Kerkoerle HLJM, Vet AJTM (1987) Radiation damage to pacemakers from radiotherapy, *PACE* 10: 538—542.
90. Venselaar JLM (1985) The effects of ionizing radiation on eight cardiac pacemakers and the influence of electromagnetic interference from two linear accelerators. *Radiotherapy and Oncology* 3: 81—87.
91. Adamec R, Haefliger JM, Killisch JP, Niederer J, Jaquet P (1982) Damaging effect of therapeutic radiation on programmable pacemakers. *PACE* 5: 146—150.
92. Maxted KJ (1984) The effect of therapeutic x- radiation on a sample of pacemaker generators. *Phys. Med. Biol.* 29: 1143—1146.
93. Hardage ML, Marbach JR, Winsor DW (1985) The pacemaker patient in the therapeutic and diagnostic device environment, pp. 857—873 in Barold SS (ed.), *Modern Cardiac Pacing.*
94. Fetter J, Aram G, Holmes DR, Gray JE, Hayes DL (1984) The effects of nuclear magnetic resonance imagers on external and implantable pulse generators. *PACE* 7: 720—727.
95. Holmes DR, Hayes DL, Gray JE, Merideth J (1986) The effects of magnetic resonance imaging on implantable pulse generators. *PACE* 9: 360—370.
96. Chaussy C, Schmiedt E (1983) Shock wave treatment for stones in the upper urinary tract. *Urol Clin North Am* 10: 743—751.
97. Langberg J, Abber J, Thuroff JW, Griffin JC (1987) The effects of extracorporeal shock wave lithotripsy on pacemaker function. *PACE* 10: 1142—1146.
98. Cooper D, Wilkoff BW, Masterson M, Castle L, Belco K, Simmons T, Morant V, Streem S, Maloney J (1988) Effects of extracorporeal shock wave lithotripsy on cardiac pacemakers and its safety in patients with implanted cardiac pacemakers. *PACE* 11: 1607—1616.

9. Antitachycardia Pacing

LUC JORDAENS, PATRICK VERTONGEN, ETIENNE VAN WASSENHOVE & DENIS L. CLEMENT

INTRODUCTION

Non-pharmacologic treatment of arrhythmias is very attractive for various reasons. Antiarrhythmic drug therapy has several disadvantages, as side effects are common and the therapy is sometimes worse than the disease. A lot of attention has recently been given to proarrhythmic side effects of antiarrhythmic drugs [1]. Furthermore, when we are dealing with a disease with paroxysmal behaviour, drug therapy poses additional problems. [2, 3]. It is a general experience that it is very difficult to convince young patients to remain compliant to the prescribed therapy. When we are facing elderly patients with structural heart disease, and suffering from potentially lethal arrhythmias, the problem of drug compliance becomes even more important [4].

Therefore, antitachycardia devices can offer a very valuable alternative to both categories of patients [5]. However, due to the recent developments of interventional techniques (surgery, catheter ablation of AV-node, and circus movement tachycardia), the indications for permanent antitachycardia pacing seem to decrease or at least will decrease [6—8]. In this overview, we will try to summarize the most important literature in this field and compare it to our own experience to redefine the possible indications for antitachycardia pacing.

PREVENTIVE PACING

Prevention of bradycardia alone can be sufficient to control tachyarrhythmias in general. Bradycardia-dependent extrasystoles can induce tachycardia and can be eliminated by pacing at conventional rates, or at faster rates ('overdrive'). In this way, the repolarization of cardiac tissue becomes more homogeneous, and reentry can be prevented. A typical example is the prevention of torsades de pointes by pacing at rates of 80 to 100 beats/min

E. Andries, P. Brugada & R. Stroobandt (eds.), How to face 'the faces' of cardiac pacing, 183—190.

[9]. Another example is preventive atrial pacing in vagally induced atrial fibrillation [10]. Preventive pacing is used a lot more than is generally thought [11]. The incidence of symptomatic, sustained atrial arrhythmias is indeed very low in our patients who were treated with physiologic pacemakers, as compared to other series [12, 13].

Other basic principles of preventive pacing are the delivery of subthreshold stimuli in the refractory period ('conditioning'), and stimulation in the ventricle at the site of early activation to prevent ventricular tachycardia [14, 15]. However, both principles were never extensively applied in clinical cardiology [16].

THERAPEUTIC PACING

Pacemaker therapy was introduced for treatment of heart block and other bradyarrhythmias. However, it became clear from the first studies using programmed electrical stimulation, that it would be possible to use pacing for tachycardia termination [5, 17]. Its application for ventricular tachycardia proved to be very useful in the setting of emergency treatment of ventricular tachycardia, as cardioversion can be avoided in patients with recurrent tachycardia. Termination of tachycardia is achieved by making a part of the arrhythmia circuit refractory by stimulation. This can be done with a single stimulus, or by multiple stimuli. Whether the tachycardia is terminated depends on multiple factors, as the position of the electrode, the characteristics of the tissue involved in the reentry circuit, the rate of the tachycardia and the timing of the stimulation in relation to the excitable gap [18]. Many algorithms were incorporated in implantable devices which can be programmed to terminate tachycardias automatically or after an activation procedure. This can usually be done by magnet application, or by programming. This method is safer when it is judged that pacing could accelerate an arrhythmia. This can be life-threatening if it occurs in the ventricle. Recently, an excellent review was published describing most pacemaker algorithms [18]. Single extrastimuli are only seldom of clinical value. Burst pacing (multiple stimuli, usually a limited number) is more effective, at least for faster tachycardias [19]. However, they are often not suited for tachycardias with a changing rate. Intelligent antitachycardia modes are 'concertina' pacing, as possible with PASAR II (Telectronics), and adaptive burst pacing (Medtronic) [20, 21]. The concertina mode is an adaptive coupled burst, limited to 7 stimuli (PASAR II), or to 8 stimuli (as tested with Intertach, Intermedics) [20]. It proved to be very effective in the termination of tachycardia. Its pacing interval is usually programmed from 65 to 80% of the tachycardia interval. Adaptive coupling with an increasing number of stimuli seems promising as well [22].

PACING FOR TERMINATION OF SUPRAVENTRICULAR TACHYCARDIA

If antitachycardia pacing is considered, extensive preoperative testing is required. This includes the proof of effective termination of the tachycardia in various conditions and positions. The electrophysiologic mechanism should be compatible with the proposed pacing configuration and pacing algorithm. For example, some groups consider an effective refractory period of 300 ms of the accessory pathway as the limit for antitachycardia pacing in the atrium, for patients with the WPW syndrome [18]. One has also to consider the possibility of fast conduction through the AV-node, during fast stimulation. A clinical contra-indication is the occurrence of syncope before an arrhythmia is terminated by pacing [23]. Accurate and fast recognition of tachycardia is therefore crucial for good antitachycardia pacing. Various algorithms for tachycardia recognition are now available. They can combine rate, sudden rate increase, sustained high rate, and rate stability [24]. The maximal heart rate during exercise should be known in order to avoid pacing during sinus tachycardia.

Long-term results with antitachycardia pacing are good. For supraventricular tachycardia, continued efficacy for 2 years was reported in 85% of the patients [25]. However, this figure seems to decrease in later years to 68%. This is probably due to the occurrence of atrial flutter and/or fibrillation. Even intelligent antitachycardia pacing, as made possible with actual multiprogrammable pacemakers, cannot avoid that atrial flutter will be induced in some patients. However, this complication seems to be often associated with undersensing [24]. This is in contrast with other nonpharmacological solutions for arrhythmias, as surgery and ablation: the incidence of atrial fibrillation is decreased after interruption of atrioventricular accessory pathways.

Two out of 30 patients were admitted to the hospital because of atrial fibrillation in the series reported by Den Dulk *et al* [26]. The incidence was lower in the series reported by Kappenberger [27]. In our experience, the incidence was initially low, but atrial fibrillation impaired antitachycardia pacing during long-term follow-up in 5 of 17 patients [28]. Antiarrhythmic drug therapy is still necessary in 42 to 49% of patients, to avoid atrial fibrillation, to slow tachycardia, or to prevent frequent recurrences.

PACING FOR TERMINATION OF VENTRICULAR TACHYCARDIA

Pacing for the termination of ventricular tachycardia is facing other problems: acceleration can lead to ventricular fibrillation. This has been described in patients in whom no acceleration was observed during extensive preoperative testing. In an overview by Fisher, 15 out of 192 patients had serious problems due to recurrent arrhythmias. Four patients in a series of 20 had sudden death, in spite of an implanted antitachycardia pacemaker [25]. New pacing techniques are under evaluation, and they seem to cause less acceleration than the classical techniques [29]. However, since the development of the implantable cardioverter-defibrillator (ICD) by Mirowski, it became accepted that no automatic antitachycardia pacing in the ventricle should be provided without the back-up of such a device [30].

We did use permanent pacing in the management of patients with ventricular tachycardia/fibrillation. However, almost all patients suffered primarily from bradycardia, due to their underlying disease, or due to antiarrhythmic drugs or beta-blockers. In Figure 1, our series of patients with ventricular tachycardia is presented. Seventeen required bradycardia support before they had their first episode for tachycardia, or after they were admitted because of this arrhythmia. A total number of 4 patients, included in this group, underwent ablation of the AV-node, to facilitate control of atrial (and ventricular!) arrhythmias [31]. Only 2 patients received an antitachycardia system, provided exclusively for termination of tachycardia. In one of these patients, an ICD was implanted as well. In general, physiological pacing was preferred to antitachycardia pacing whenever it was possible. Nevertheless, even in patients with a rate-responsive pacemaker, or with a DDD, it was possible to terminate ventricular tachycardia after admission in the emergency department (Figures 2 and 3). DDD pacing improved the quality of life in two of our patients with an ICD, without interference of both systems. Another advantage in patients with an implanted antitachycardia unit is that it can be used as E.P.-lab, avoiding repeated invasive studies. This proved to be very useful in some of our patients, even when an ICD was present as well.

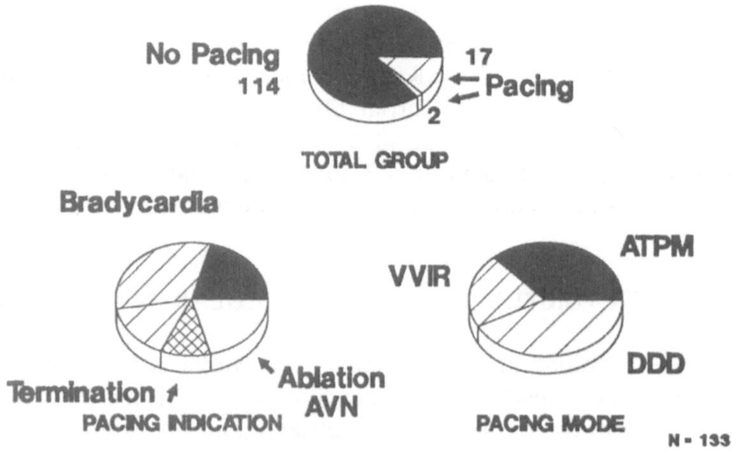

Figure 1. Use of pacemakers in a study group with ventricular tachycardia or ventricular fibrillation.

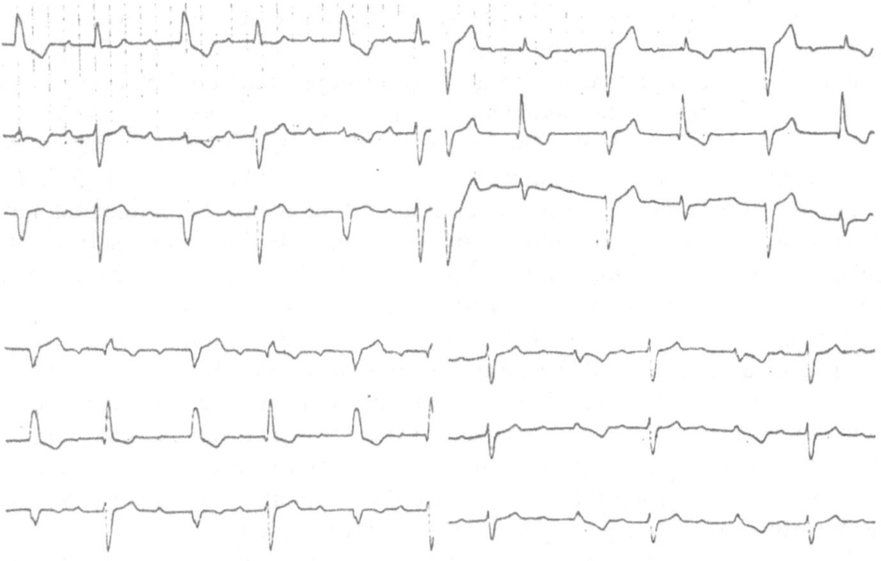

Figure 2. Alternating right and left bundle branch block with first degree heart block in a 61-year old female patient with aortic valve and coronary artery disease, complicated by recurrent sustained ventricular tachycardia.

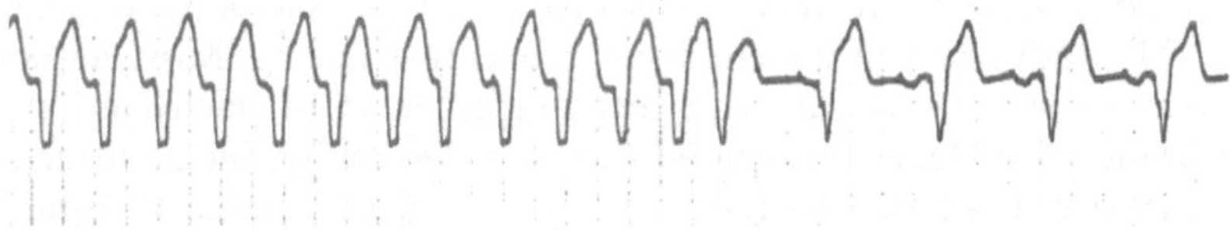

Figure 3. Same patient as in Figure 2. On admission in the emergency department, she had recurrence of ventricular tachycardia, slowed by sotalol treatment. Magnet application over the pacemaker interrupts the arrhythmia with one captured beat, and resumption of pacing at atrial and ventricular level.

HAS ANTITACHYCARDIA PACING A FUTURE?

The actual advances in ablative and surgical therapy limit the potential applications of antitachycardia pacing in supraventricular tachycardia. However, when symptomatic bradycardia is present, and pacing is necessary anyhow, antitachycardia pacing remains attractive. AV-nodal tachycardia is less often considered for surgical correction, because it is not associated with lethal arrhythmias as are some tachyarrhythmias in the presence of antegrade conducting accessory bypass tracts. Furthermore, catheter ablation of AV-nodal tachycardia is often associated with the occurrence of temporary (or longer-lasting) heart block. Therefore, an antitachycardia unit that is implanted in such a situation needs to pace in the ventricle as well [32]. This could also be helpful in other situations. Some forms of congenital heart disease are associated with important junctional arrhythmias and supraventricular tachycardia, including atrial flutter. Pacing is often necessary from the haemodynamic point of view, to provide bradycardia support. It is reported that cardiac performance increases considerably after implantation, and that the rate of attacks of tachycardia diminishes progressively after implantation [33]. This is also our experience in young adults with complex congenital heart disease. Ablative and surgical procedures are not always easy, or desirable, in this patient group. However, it has to be mentioned that it is exactly in this subgroup that we did have the problem of frequently recurrent tachycardia. Preventive pacing remains therefore attractive.

In our population of patients, paced for bradycardia, atrial flutter sometimes occurs, even without evidence of organic heart disease (5/252 with a mean follow-up of 2.5 years). The additional possibility of providing antitachycardia pacing to terminate flutter should therefore be considered for incorporation in DDD-pacemakers, in spite of reports suggesting that the need for antitachycardia features in pacemakers implanted for bradycardia is low [34].

The potential benefits of antitachycardia pacing in the ventricle without the back-up of an ICD are very limited. However, an occasional patient with a long QT syndrome will benefit from preventive pacing. Most authorities agree that some patients can have advantages of an antitachycardia pacemaker in the ventricle, as this can be used for electrophysiologic testing, and because physician-assisted termination may be an acceptable treatment for carefully selected patients [35].

The new ICD's have antitachycardia pacing available as well. However, extensive programming of antitachycardia pacing may limit the benefits for some patients: they may have syncope before termination, or they will be shocked after a long episode of tachycardia, with an increased chance of failure of shock therapy. Patients with ventricular arrhythmias and poor left ventricular function could theoretically have problems when they are paced in the ventricle. In such patients, DDD pacing is preferential to VVI pacing. No ICD has such pacing functions available. Therefore, it is unlikely that all pacemaker and ICD interactions will completely disappear in the next few years.

REFERENCES

1. Cardiac Arrhythmias Suppression Trial (CAST) (1989) Preliminary report: effect of encainide and flecainide on mortality in a randomized trial of arrhythmia suppression after myocardial infarction. *N Engl J Med* 321: 406—412.
2. Margolis B, DeSilva RA, Lown B (1980) Episodic drug treatment in the management of paroxysmal arrhythmias. *Am J Cardiol* 45: 621—626.
3. Pritchett EL, Smith MS, McCarthy EA, Lee KL (1984) The spontaneous occurrence of paroxysmal supraventricular tachycardia. *Circulation* 70: 1—6.
4. Gleed KJ, Hopson R, Martins JB (1991) Long-term follow-up of patients with inducible sustained monomorphic ventricular tachycardia and heart disease (abstract). *J Am Coll Cardiol* 17: 31A.
5. Wellens HJJ, Bär FW, Gorgels AP, Farré Muncharaz J (1978) Electrical management of arrhythmias with emphasis on the tachycardias. *Am J Cardiol* 41: 1025—1034.
6. Gallagher JJ, Svenson RH, Kasell JH, *et al.* (1982) Catheter technique for closed-chest ablation of the atrioventricular conduction system. *N Engl J Med* 306: 194—200.
7. Haissaguerre M, Warin JF, Lemetayer P, Saoudi N, Guillem JP, Blanchot P (1989) Closed-chest ablation of retrograde conduction in patients with atrioventricular nodal reentrant tachycardia. *N Engl J Med* 320: 426—433.
8. Cox JL (1983) The surgical management of cardiac arrhythmias, pp. 1552—1584 in Sabiston DC, Spencer FC (eds) *Gibson's Surgery of the Chest.* Philadelphia: WB Saunders.
9. Khan MM, Logan KR, McComb JM, Adgey AAJ (1981) Management of recurrent ventricular tachyarrhythmias associated with Q-T prolongation. *Am J Cardiol* 47: 1301—1308.
10. Coumel P, Friocourt P, Mugica J, Attuel P, Leclercq JF (1983) Long-term prevention of vagal atrial arrhythmias by atrial pacing at 90/minute: Experience with 6 cases. *PACE* 6: 552—560.
11. Rosenqvist M, Brandt J, Schüller H (1988) Long-term pacing in sinus node disease: effects of stimulation mode on cardiovascular morbidity and mortality. *Am Heart J* 116: 16—22.
12. Jordaens L, Robbens E, Van Wassenhove E, Clement DL (1989) Incidence of arrhythmias after atrial or dual-chamber pacemaker implantation. *Eur Heart J* 10: 102—107.
13. Snoeck J, Decoster H, Verherstraeten M, *et al.* (1990) Evolution of P-wave characteristics postpacemaker implantation. *PACE* 13: 2091—2095.
14. Prystowsky EN, Zipes DP (1983) Inhibition in the human heart. *Circulation* 68: 707—713.
15. Marchlinski FE, Buxton AE, Miller JM, Josephson ME (1987) Prevention of ventricular tachycardia induction during right ventricular programmed electrical stimulation by high current strength pacing at the site of origin. *Circulation* 76: 332—342.
16. Kuck KH, Kunze KP, Schlüter M, Bleifeld W (1984) Tachycardia prevention by programmed stimulation. *Am J Cardiol* 54: 550—554.
17. Wellens HJ, Schuilenburg RM, Durrer D (1972) Electrical stimulation of the heart in patients with ventricular tachycardia. *Circulation* 46: 216—226.
18. De Belder MA, Malik M, Ward DE, Camm AJ (1990) Pacing modalities for tachycardia termination. *PACE* 13: 231—248.
19. Naccarelli GV, Zipes DP, Rahilly GT, Heger JJ, Prystowsky EN (1983) Influence of tachycardia cycle length and antiarrhythmic drugs on pacing termination and acceleration of ventricular tachycardia. *Am Heart J* 105: 1—5.
20. Spurrell RAJ, Nathan AW, Camm AJ (1984) Clinical experience with implantable scanning reversion pacemakers. *PACE* 7: 1296—1299.
21. Charos GS, Haffajee CI, Gold RL, Bishop RL, Berkovits BV, Alpert JS (1986) A theoretically and practically more effective method for interruption of ventricular tachycardia: selfadapting autodecremental overdrive pacing. *Circulation* 73: 309—315.

22. Den Dulk K, Kersschot IE, Brugada P, Wellens HJJ (1988) Is there an universal antitachycardia pacing mode? *Am J Cardiol* 57: 950—955.
23. Curry PV (1979) The hemodynamic and electrophysiological effects of paroxysmal tachycardia, pp. 364—381 in Narula O (ed.), *Cardiac Arrhythmias*, Electrophysiology, Diagnosis and Management. Baltimore: Williams and Wilkins.
24. Fromer M, Gloor H, Kus T, Shenasa M (1990) Clinical experience with a new software-based antitachycardia pacemaker for recurrent supraventricular and ventricular tachycardias. *PACE* 13: 890—899.
25. Fisher JD, Johnston DR, Furman S, Mercando AD, Kim SG (1987) Long-term efficacy of antitachycardia pacing for supraventricular and ventricular tachycardias. *Am J Cardiol* 60: 1311—1316.
26. Den Dulk K, Brugada P, Smeets JLRM, Wellens HJJ (1990) Long-term antitachycardia pacing experience for supraventricular tachycardia. *PACE* 13: 1020—1030.
27. Kappenberger L, Vallin H, Sowton E (1989) Multicenter long-term results of antitachycardia pacing for supraventricular tachycardias. *Am J Cardiol* 64: 191—193.
28. Jordaens L, Van Wassenhove E, Clement DL (1986) An implantable anti-tachycardia pacemaker with back-up pacing and scanning burst mode. *Eur Heart J* 7: 61—66.
29. Gardner JM, Waxman HL, Buxton AE, Cain ME, Josephson ME (1982) Termination of ventricular tachycardia: Evaluation of a new pacing method. *Am J Cardiol* 50: 1338—1345.
30. Mirowski M (1985) The automatic implantable cardioverter-defibrillator: an overview. *J Am Coll Cardiol* 6: 461—466.
31. Critelli G, Scherillo M, Monda V, D'Ascia C, Musumeci S, Antignano A (1986) Transvenous catheter ablation of the His bundle in ventricular tachycardia. *Am Heart J* 111: 1106—1112.
32. Zipes DP, Prystowsky EN, Miles WM, Heger JJ (1984) Initial experience with Symbios model 7008 pacemaker. *PACE* 7: 1301—1305.
33. Gillette PC, Zeigler V, Kratz J, Oslizok P (1991) Cardiac pacing in children and young adults, pp. 406—412 in Horowitz LN (ed.), *Current Management of Arrhythmias.* Philadelphia: Marcel Decker.
34. Den Dulk K, van Wylick ARJM, Kersemakers JGM, Wellens HJJ (1985) Do all pacemakers need both antibradycardia and antitachycardia pacing features? *Am J Cardiol* 55: 593—594.
35. Rosenthal ME, Josephson ME (1990) Current status of antitachycardia devices. *Circulation* 82: 1889—1899.

10. Electrical treatment of tachycardias

JEAN FRANÇOIS LECLERCQ, ISABELLE DENJOY,
ANTOINE LEENHARDT & PHILIPPE COUMEL

INTRODUCTION

The vast majority of patients with paroxysmal tachycardias, namely supraventricular tachyarrhythmias (SVT), respond to antiarrhythmic drugs. However, since SVT usually have a good prognosis, problems occurring during long-term follow-up, especially the adverse effects of drugs, may lead to the use, in some cases, of nonpharmacological therapies. Finally, some SVT are definitively resistant to any drug combination, and one of the 3 possible nonpharmacological therapies (surgery, ablation, or antitachycardia pacing) has to be chosen. Patients with ventricular tachyarrythmias may also be candidates for a nonpharmacological treatment, surgery, or ablation in the case of ventricular tachycardia (VT), or electrical devices in the case of recurrence, despite antiarrhythmic drug combinations.

1. ELECTRICAL TREATMENTS OF SVT

1.1. Ablation in SVT: Advantages and limitations

The advantages of ablation in SVT are evident: it is a curative technique, simpler than surgery, with much less potential morbidity and a shorter recovery period.

But this ideal cannot easily be reached in all cases. Some limitations are evident:

— in atrial fibrillation, no curative ablation is feasible, but only palliative ablation of the AV conduction with a mandatory VVI (preferably VVI,R) pacemaker. This therapy could also be applied to all types of atrial arrhythmias.

— with the currently used technique of electrical fulguration, the generated barotrauma is dangerous, in some cases leading to an atrial rupture. The initial series of ablation of left posteroseptal accessory pathways into the

E. Andries, P. Brugada & R. Stroobandt (eds.), How to face 'the faces' of cardiac pacing, 191—207.

coronary sinus were particularly demonstrative [1—3]. In the case of atrial ablation (mainly in common atrial flutter), the risk of atrial rupture seems especially high, even with a reduced energy. We had a personal experience of an immediate hemopericardium leading to an extremely urgent successful surgical intervention in this setting. That is why Dr Chauvin, one of the pioneers of this type of ablation in Strasbourg [4], has stopped ablation of atrial flutter for the moment. Nevertheless, some groups have been able to publish interesting results in atrial flutter [5], or atrial foci [6]. However, this technique of atrial ablation must be considered as still investigational, and major improvements are needed before its generalization in common practice.

In direct fulguration of accessory pathways, the most common indication of the technique at present, an hematoma or a rupture of the AV groove, may also occur, even with direct ablation, avoiding shocks within the coronary sinus. Some groups, especially in France, now have relatively wide experience of this technique, and during a recent meeting of the French working group on arrhythmias, we realized that we had a significant incidence of serious complications and some deaths in all groups. A meta-analysis of the series of Bordeaux, Paris (Jean Rostand and Lariboisière), and Nancy evidenced 5 deaths out of 280 patients (1.8%) and 6 pericardial effusions with tamponade requiring immediate surgical intervention (2%).

For these reasons, some groups attempted to develop new techniques of ablation, avoiding the risk of cardiac rupture: the only technique presently able to give some success is the radiofrequency desiccation. Contrarily to fulguration, this technique does not produce arcing or explosive gas formation. However, it is still investigational, since the success rate is low with the current material, even in AV junctional ablation [7]. Its application to other forms of SVT still remains limited to some groups, although with inconsistent results, even if the success rate of the procedure is gradually increasing with time [8, 9]

— ablation of SVT is difficult, because the endocardial mapping is often time-consuming (several hours) and the criteria of an accurate location for ablation is not always perfectly defined. The impressive and excellent results of the Bordeaux group [10—12] are not easily reproducible in other centers [13] because the technique requires the use of top quality material and a very well trained team of electrophysiologists.
— ablation of the so-called AV nodal reciprocating tachycardia is also feasible [11], but with a risk of AV block leading to a pacemaker implant, even when used by an experienced electrophysiologist.
— ablation of the accessory pathway responsible for the permanent form of AV junctional reciprocating tachycardias, which is an AV left posteroseptal accessory pathway, is similar to the WPW ablation procedure. Initially performed in the coronary sinus [14], it is perfectly feasible in the lower part of the right atrium [12, 15].

The main alternative to ablation, especially in the Wolff—Parkinson—White (WPW) syndrome, is surgery. This realizes the definitive cure of the arrhythmia with an excellent quality of life in the follow-up. The overall success rate is high when performed by a well-trained surgeon [16, 17], because the surgical techniques are now relatively well known. As an example, Table 1 display our personal results from 1983 to 1990. However, the limitations of WPW surgery are obvious:

— peri-operative mortality is fortunately very low, but cannot be neglected in patients with SVT: even in the case of atrial fibrillation with fast ventricular response, the spontaneous motality of the patients is low with medical therapy and, ideally, surgery should have a mortality rate of zero, which is almost impossible.
— operative morbidity is also low, but cannot be eliminated. The risk of AV block in the surgical cure of septal pathways is acceptable, but must not be underestimated. The risk of mediastinitis or other complications of all cardiac surgery cannot be ruled out.
— some SVT are not or less accessible to surgery. Atrial fibrillation cannot be considered as an indication for surgery, nor for ablation. Surgery of atrial ectopic focus, or atrial flutter is difficult and not routinely performed. More recently, a surgical curative attempt in AV nodal reentry was performed with good results [18, 19]. However, this experience is limited to few surgeons, and practical surgery in SVT still remains limited to the WPW syndrome.

Even in WPW syndrome surgery, the success rate is obviously not 100%, mainly because of the incidence of multiple accessory pathways. In a surgical series, biased because it preferentially includes high-risk or highly resistant arrhythmias, the incidence of such multiple pathways is relatively high (24% in our series).

Unfortunately, electrophysiological mapping during pre-operative study, as well as pre-operative epicardial mapping, are not able to provide an accurate diagnosis in all cases. In that situation, a complete or partial failure of the WPW surgery is unavoidable. That point remains a major

Table 1. Results of WPW surgery according to accessory pathway localization

Localization	*n*	Success	Failure	Death
Left lateral	32	32	0	0
Left posteroseptal	12	7	4	1
Right posteroseptal	15	15	0	0
Right lateral	7	7	0	0
Right anteroseptal	1	1	0	0
Multiple pathways	21	13	7	1
Total	88	75 (85%)	11 (12.5%)	2 (2%)

problem for all groups, and indicates that a second operation may be necessary, preferably the day after the first procedure. That is why ablative techniques represent a new and important way in the treatment of SVT: it is more difficult than surgery, with a lower success rate, but there are no limitations on the number of attempts. However, it supposes that the risks of mortality and/or morbidity are especially low, which was not the case in 1990. Nevertheless, it is obvious that ablative techniques represent the future and surgery the past in the curative treatment of SVT. Major improvements in the ablation technique will result from the development of appropriate materials, especially catheters able to both record anomalous pathway electrical activity [20] and to deliver ablative electrical energy. The design of special catheters is critical for fulguration [21]: many commercially available catheters cannot be used for electrical ablation using high energy. Figure 1 shows an example of such a catheter, not able to support a modest energy (100 J), usually required for the ablation of the majority of tissues when conventional defibrillators are used. In addition, it has been underlined by the Bordeaux group that recording the Kent bundle activity is a good marker for success in WPW ablation [10, 12]. Unfortunately, such recording is not consistently possible, even after a time-consuming procedure when conventional catheters are used. Major improvements of the ablation technique are

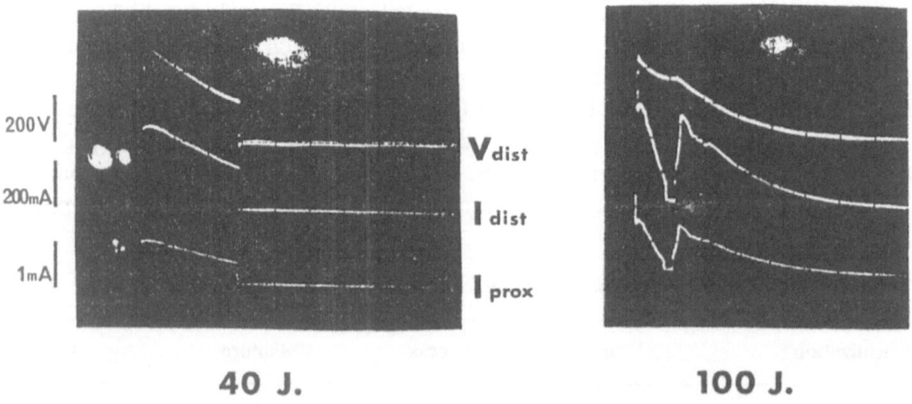

Figure 1. Recording of voltage and current curves during fulguration DC shocks delivered into a commercial bipolar catheter- electrode, using a distal electrode as the cathode and a cutaneous path as the anode. With 40 J of stored energy (left), a regular decrease of both current and voltage are observed in the distal electrode, with minimal shunt-effect in the proximal one. Increasing stored energy to 100 J induced an abrupt fall clearly visible on the current curve: this catheter is not able to support this charge. Testing the catheters before ablation is mandatory, and designing special catheters for ablation is desirable.

then necessary before considering it as a first-choice therapy in refractory SVT.

1.2. Antitachycardia pacing in SVT: advantages and limitations

The main advantage of antitachycardia pacing is its relative simplicity compared to ablation or surgery, with an absence of any significant risk of mortality, and low morbidity. For that reason, antitachycardia pacing was developed before other nonpharmacological therapies [22, 23].

The main limitation is that it is a palliative and not a curative therapy. Usually, attacks of SVT persist for a long time and often for all of the patient's life. Several antitachycardia devices may be then necessary for the same patient.

Two types of antitachycardia pacing may be used: preventive and/or curative.

Prevention of some SVT can be reached by simple AAI (or DDD) pacing in some cases. It is well known, for example, that AV reciprocating tachycardias may be initiated by junctional escape beats leading to a retrograde V-A conduction. But the main application of the preventive atrial pacing are paroxysmal atrial arrhythmias: atrial fibrillation or flutter. In these patients, atrial pacing itself is able to reduce the incidence of atrial arrhythmias. It has been evidenced, for example, that AAI pacing is superior to VVI pacing in the sick sinus syndrome in preventing the appearance of chronic atrial fibrillation [24]. Similarly, the incidence of systemic embolism is reduced and survival increases [24, 25]. This latter point is favoured for a reduction of the episodes of paroxysmal atrial arrhythmias. Finally, even in patients without sinus-node dysfunction and having a vagally-dependent or a bradycardia-dependent arrhythmia, fast atrial pacing is useful as a preventive treatment of paroxysmal atrial arrhythmias [26] when drug therapy is ineffective. It has been postulated that some specific algorithms of pacing which are able to 'regularize' atrial rhythms (flywheel, rate-smoothing, AAIR modes), should be even more efficient in preventing atrial arrhythmias, but that has never been demonstrated by clinical trials.

However, in organized SVT, antitachycardia pacing is usually used as a curative treatment, i.e. to stop SVT and not to prevent it even if back-up AAI pacing is often present, and can be used in some cases as a preventive combined therapy. Several modes of antitachycardia pacing are commonly available: dual-demand AAI [27], VVI or DDD pacing, atrial or ventricular extrastimulations with scanning of the tachycardia's cycle [28, 29] and, finally, atrial bursts, which are commonly efficient in the majority of regular SVT. Several modes of the variation of coupling and cycle intervals of these bursts could be designed in order to approach the ideal antitachycardia pacing mode, i.e. that able to stop SVT after a minimum number of attempts [30, 21]. Figures 2 and 3 represent two examples of successful modes of

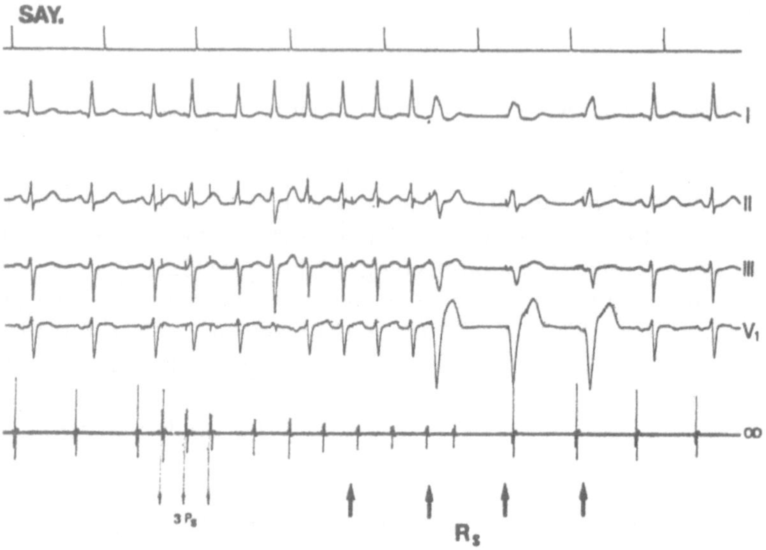

Figure 2. Patient with AV reciprocating tachycardia (latent accessory pathway) implanted with a dual-demand VVI pacemaker. The tachycardia was induced by 3 atrial extrastimulations. Based only on the rate criterion, SVT is recognized after 4 beats and the pacemaker switched to the VOO mode. The first stimulus falling outside the refractory period stops the tachycardia, anticipating the following atrial depolarization which is blocked in the AV node. Back-up VVI pacing is then used.

antitachycardia pacing (dual-demand VVI pacing and ventricular extrastimulations with automatic scanning) in two cases of reciprocating AV tachycardia.

An important limitation of atrial antitachycardia pacing is the possibility of inducing an atrial fibrillation. This problem is frequently encountered in relatively old patients vulnerable to frequent bursts of fast atrial pacing [32]. In patients with the WPW syndrome, having a good antegrade conduction through the accessory pathway, atrial bursts, and probably almost all form of antitachycardia pacing, must be prohibited.

The main difficulty in antitachycardia pacing is probably the accurate detection of SVT by the device and, more precisely, the differentiation between SVT and sinus tachycardia [33, 34]. Even with sophisticated algorithms such as the sudden onset or rate stability of the tachycardia, an important sinus tachycardia in young patients may perfectly mimic SVT and trigger the antitachycardia device. In that situation there are two possibilities: prescribe some drugs for slowing sinus rhythm such as beta-blockers or use the manual antitachycardia mode which is available in many devices. If the

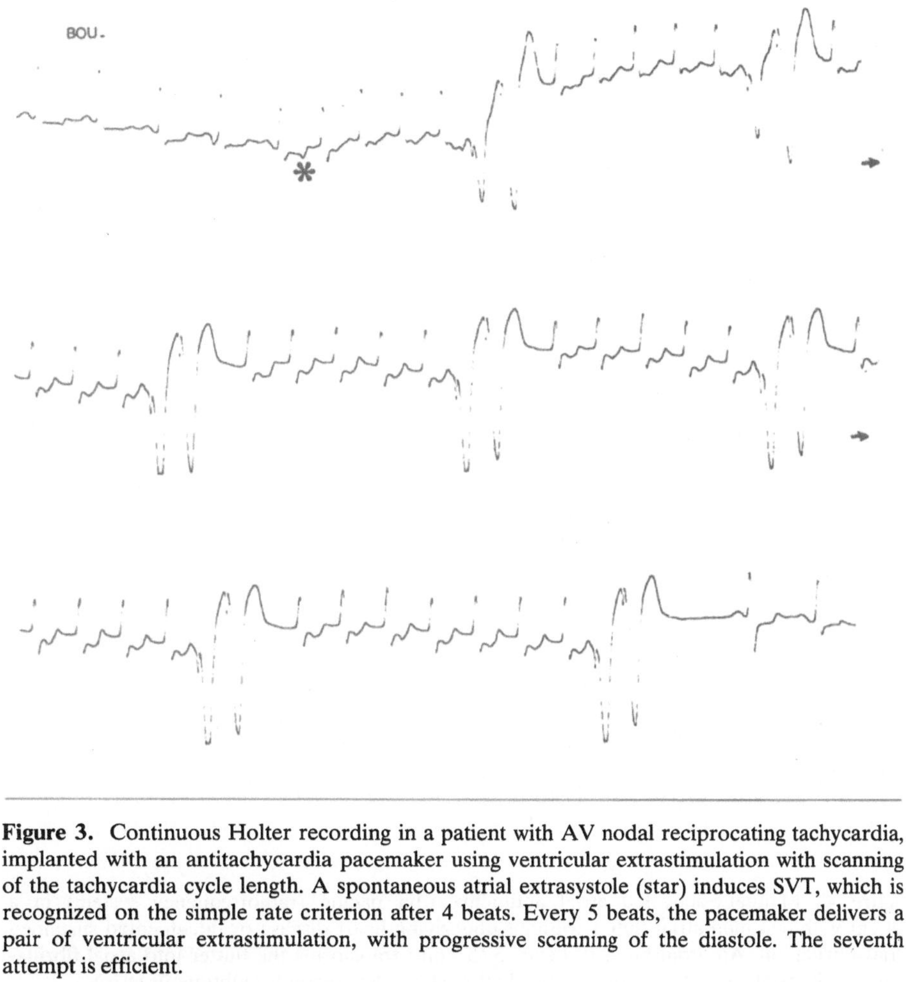

Figure 3. Continuous Holter recording in a patient with AV nodal reciprocating tachycardia, implanted with an antitachycardia pacemaker using ventricular extrastimulation with scanning of the tachycardia cycle length. A spontaneous atrial extrasystole (star) induces SVT, which is recognized on the simple rate criterion after 4 beats. Every 5 beats, the pacemaker delivers a pair of ventricular extrastimulation, with progressive scanning of the diastole. The seventh attempt is efficient.

tachycardia episodes are sporadic, the manual mode is perfectly suitable. When they occur almost daily, a small dosage of beta-blockers seems preferable. Figure 4 shows the efficacy of manually-activated atrial bursts in a patient with common atrial flutter.

Finally, there are difficulties which are not related to the pacemaker itself, but rather to the disease: the characteristics of SVT in an individual patient may change with the situation and with time. It has been well demonstrated that the tachycardia cycle length of the same SVT may change from one episode to another, depending on the circumstances and then on the status of the vago-sympathetic balance [35]. In addition, the type of SVT, as well as the mode of spontaneous initiation, may change in the same patient. All these

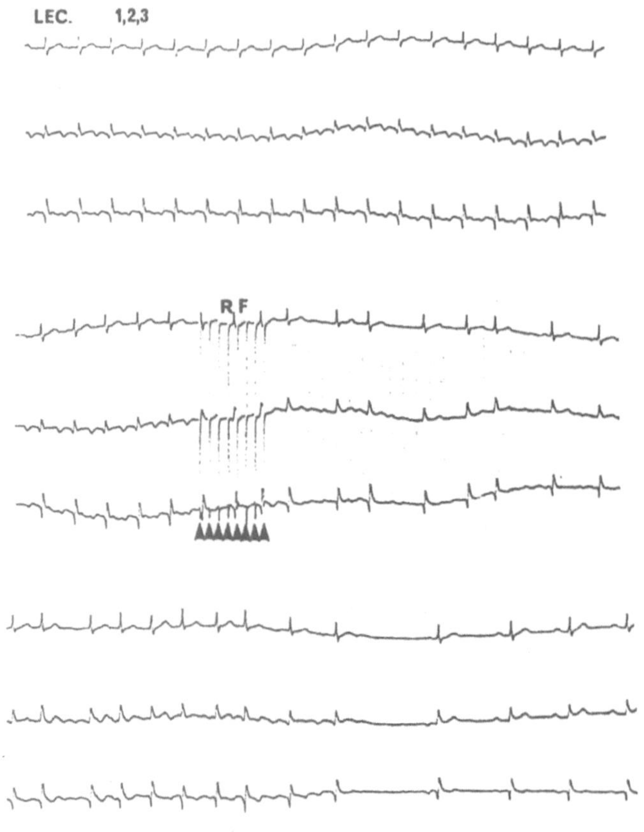

Figure 4. Manually-activated atrial antitachycardia pacing (radiofrequency system) in a patient with common atrial flutter, without underlying heart disease or documented sustained atrial fibrillation. An atrial burst (8 beats, 375 bpm) transforms the flutter into atrial fibrillation, which spontaneously reverts to sinus rhythm after a few seconds (continuous strip).

variations require changes in the antitachycardia pacing mode. That is why several changes are usually needed in these patients after implantation. During long-term follow-up, antitachycardia pacing is often stopped after several years and a curative technique (surgery or ablation) has to be performed [36].

This latter point illustrates the main difference between antitachycardia pacing and ablation: the first technique is palliative and often temporarily chosen, because the second one is curative but has not been sufficiently developed at present.

The choice of an electrical therapy for SVT in a given patient
This choice is mainly based on the type of arrhythmia, and also on the group's personal experience and on the patient's preference.

In paroxysmal atrial fibrillation, the only definitive treatment is palliative: AV ablation and VVI,R pacemaker. In patients with vagally-induced atrial fibrillation, we often propose to first implant a DDD,R pacemaker and then we are able to obtain sufficient control of the arrhythmia in many cases (always combined with antiarrhythmic therapy), thus avoiding the AV ablation, at least for some years. His's bundle ablation is not satisfactory for the physician because it is illogical to treat one disease by creating another. However, it is satisfactory for the patient, with a high percentage of clinical success [37—38]. The small incidence of unexplained sudden death in the follow-up does not seem clearly related to the procedure and the technique could be considered as safe and efficient. Some refinements, such as the incomplete ablation of the AV nodal conduction (so-called modulation) using radiofrequency ablation, may be useful but, even in the present status, His' bundle ablation represents an acceptable final treatment in many aged patients invalided by incessant palpitations, because atrial arrhythmias are often refractory to antiarrhythmic drug therapy.

In paroxysmal common atrial flutter, especially in the absence of detectable underlying heart disease, and without spontaneously documented atrial fibrillation, and antitachycardia pacing may be useful, and, at present, seems preferable to AV ablation or to atrial ablation. In the case of both atrial flutter and fibrillation, antitachycardia pacing is certainly less useful, leading almost automatically to sustained atrial fibrillation. However, the clinical tolerance of atrial fibrillation is usually better than that of flutter, mainly because of a lower ventricular rate, and even in that case, antitachycardia pacing may induce a satisfactory clinical improvement.

In WPW syndrome, a radical cure is often proposed, because it is really curative for the patient who usually has a normal underlying heart. Choosing between surgery and ablation principally depends on the personal experience of the group of rhythmologists. In the future, ablation will probably represent the best choice for the majority of patients who are refractory to drugs, or who would prefer a radical cure of the disease. Indications of antitachycardia pacing are practically limited to patients with long antegrade refractory periods of the accessory pathway and AV reciprocating tachycardia easily stopped by a few extrastimulations.

In AV nodal reciprocating tachycardias, curative treatment is not yet routinely performed. An antitachycardia pacing may represent a good alternative solution, with excellent clinical results, even if some check-up visits with reprogrammation of the device, and/or adjunctive drug therapy are frequently needed. It is obvious that in young patients, this solution should be considered as temporary: it allows the patient, waiting for an appropriate development of ablative curative techniques, to be treated without compromising the future, as with His' bundle ablation, for example.

2. ELECTRICAL TREATMENT OF VT

2.1. Ablation in VT: advantages and limitations

As for SVT, the ablative technique has as its major advantages its relative simplicity and a lower risk than surgery. In patients with refractory ventricular tachyarrhythmias, a risk of death related to the procedure is acceptable, since the spontaneous risk to the patient is high. In VT ablation, this risk is not negligible, but low [39, 40], and seems particularly related to the cardiac status. This technique seems to be safe in patients with a diseased right ventricle but who have a normal left ventricle. [39, 41].

Unfortunately, there are many limitations of VT ablation:

- technical limitations, as explained for SVT.
- limitations related to the cardiac disease and/or the arrhythmia's characteristics:
  - patients with polymorphic VT, or ventricular fibrillation (VF), without sustained monomorphic VT, are not candidates for ablation, even those with monomorphic VT initiating VF, because of the impossibility of achieving an accurate endocardial mapping.
  - patients with many different *spontaneous* monomorphic VTs (so-called pleiomorphic VT by some authors). In that case, the success rate is lower, each VT necessitating a selective mapping and ablation. It is probable that the same applies to patients with different *induced* monomorphic VTs, each representing a potential spontaneous VT. This situation is especially frequent in patients with VT due to an old myocardial infarction.
  - mapping of VT is difficult, because there is no single simple criterion of an accurate location of the catheter and, consequently, it is often hard to choose the site of ablation. It is then logical that the success rate of a single ablative procedure is relatively low. Using the initially proposed criteria for the localization of the ablation site (prematurity and pace-mapping), the results of ablation are poor in post-myocardial infarction VT [39—42], and better in arrhythmogenic right ventricular disease [39, 42]. The quantity of muscle that must be necessarily destroyed in successfully locating the catheter, may be more important in the first case, and the accurate location of the catheter may be even more critical. Moreover, no correlation was found between the quality of results and the degree of prematurity of endocardial activation [40, 42, 43].

  That is why the development of new criteria of the ideal ablation site is desirable. The concept of a slow conduction zone during VT was developed some years ago in experimental models and demonstrated in human VT [44]. Recently, it has been suggested that ablation in a slow conduction zone participating in the VT circuit may have better results than ablation based on classical mapping criteria [45]. Delay to ventricular capture after stimulus, elevated pacing threshold, concealed entrainment of the VT during pacing, and mid-diastolic endocardial potential are the main 'new' criteria proposed for the accurate identification of the ablation site. These findings have been observed in post-myocardial infarction VT. They could apply to right ventricular disease, as proved by the personal observations showed in Figures 5 to 7. In that case of refractory VT due to arrhythmogenic right ventricular dysplasia, several slow conduction zones could be identified, but only one seemed to be able to participate in the VT reentry circuit, the other corresponding to 'dead-end' zones in which no entrainment of the VT was possible. Fulguration in the identified 'circuit-connected' slow conduction zone was perfectly efficient during the 3 years of follow-up without drug therapy. The same difficulty in identifying participating and nonparticipating slow conduction zones can be observed in other types of VT, explaining that determining the site to ablate often remains difficult in clinical practice.

— A particular case is represented by bundle-branch reentry: In that situation, the localization of the ablation site is easy, since it is a branch of the His bundle, usually the right bundle branch [46, 47]. Unfortunately, this kind of VT is rare, at least in post-myocardial infarction VT.

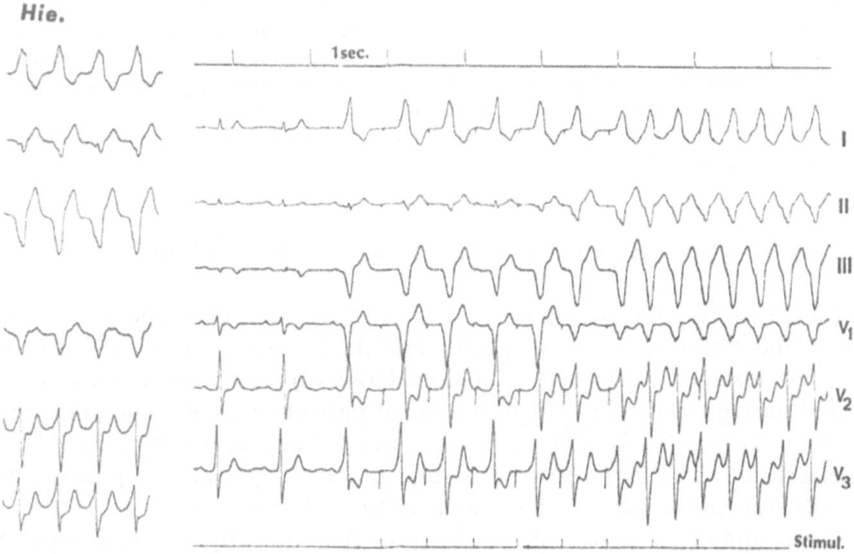

Figure 5. Patient with arrhythmogenic right ventricular dysplasia and refractory-sustained VT with LBBB and LAD pattern (left) originating from the right inferior wall. In the chosen ablation site, a slow conduction zone is demonstrated by the long delay between the stimulus artefact and ventricular response (right). The pacing threshold was high: 18 mA. Note the perfect similarity between paced beats and VT beats.

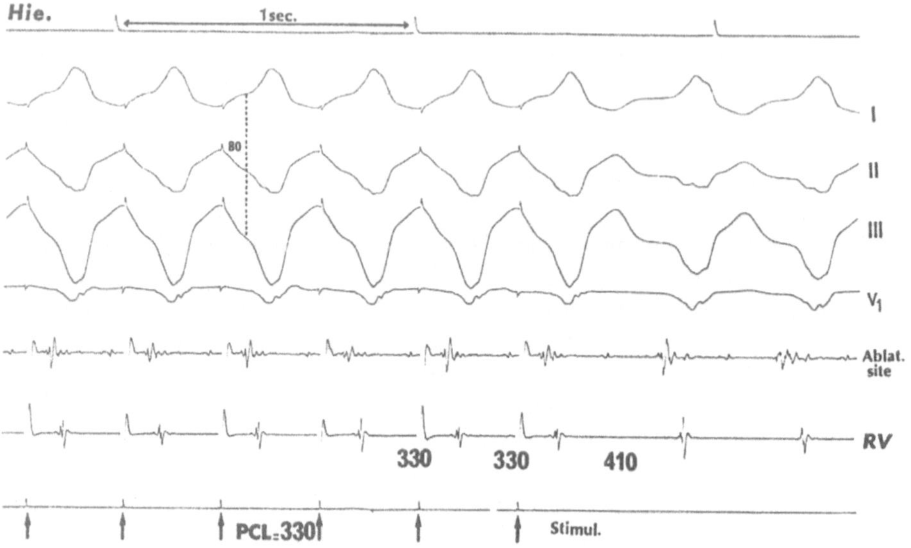

Figure 6. Same patient as in Figure 5. Ventricular pacing (cycle length = 330 ms) during VT in the vicinity of the chosen ablation site. A slow conduction is present, evidenced by the delay of 80 ms between stimulus artefact and ventricular response. However, after the last stimulation, VT resumes with its initial cycle length (410 ms). Note the slight differences between paced beats and VT beats: paced beats are fusion beats. Recording at the chosen ablation site evidenced the mid-diastolic potential clearly separated from the rest of the ventricular activity, and related to the preceding QRS beat.

2.2. Antitachycardia pacing in VT: advantages and limitations

Using antitachycardia pacing in VT is a relatively old idea. The main advantage of this technique is the simplicity, and the low risk of the implant procedure, compared to surgery or ablation. However, it has been well demonstrated during electrophysiologic studies of patients with VT that (1) efficacy of pacing to stop VT is relatively low, compared to that observed in SVT, (2) acceleration of VT into VF is a significant risk, especially in patients with underlying heart disease, i.e. the vast majority of VT patients.

A first and most simple solution is to restrict antitachycardia pacing of VT to the manual mode by means of a radiofrequency system, or DDT mode [48], activated by the physician and not by the patient himself. In that situation, which is advocated by our group as well as by all French groups, the usefulness of antitachycardia pacing is confined to stopping the VT of the patient in the emergency room, thus avoiding the use of other electrical treatment.

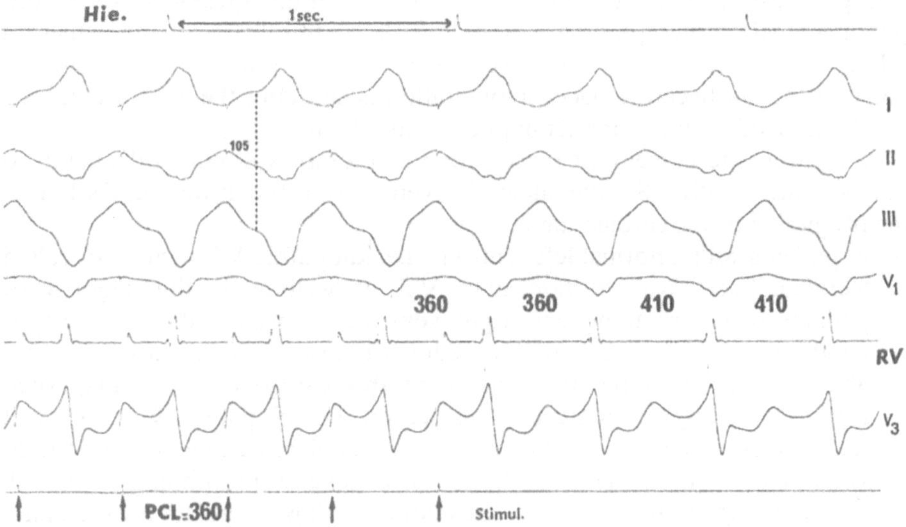

Figure 7. Same patient as in Figures 5 and 6. Pacing (cycle length = 360 ms) at the chosen ablation site. The slow conduction zone induces an even longer delay (105 ms) between the stimulus artefact and ventricular response than in the preceding site. After the last paced beat, note (1) the return cycle to VT identical to pacing cycle length (360 ms and not 410 ms), contrarily to that observed in Figure 6, and (2) the perfect similarity between paced beats and VT beats, demonstrating the concealed entrainment of the VT.

Other groups studied comparatively different modes of antitachycardia pacing in terms of efficacy and safety. Efficacy of underdrive dual-demand pacing seems very low. Automatic overdrive pacing is not satisfactory in VT, contrarily to SVT, with a high incidence of failures or accelerations [36]. The success rate is 76%, but with 35% VT acceleration [49]. Diastolic scanning, with up to 3 extrastimuli, is less hazardous, but also less efficient: only 3% VT acceleration, but with a 44 to 63% rate of success [49].

Autodecrementation of the overdrive pacing seems an optimal combination of these two modes [30, 50]. However, all these groups underlined a major fact that antitachycardia pacing efficacy is closely related to the VT rate: patients with slower VT are responders, while those with faster VT have a high incidence of failures or VT acceleration [49, 50]. That is why antitachycardia pacing in VT concerns very few patients and, then, only those with a relatively good prognosis. In the majority of patients with VT, antitachycardia pacing must be combined with automatic defibrillation within the same device, as is now realized by several manufacturers.

The choice of an electrical treatment in a given patient with VT

The prognosis of patients with VT is poor, generally speaking. Some non-pharmacological therapies have a high risk of immediate death which is related to the procedure. This is true for surgery, and to a lesser extent, for ablation. Antitachycardia pacer implantation is safe, but the risk of death by acceleration of VT into VF after implantation is high.

Moreover, the efficacy of these different therapies is not 100%, and no one is really curative. So, the choice is even more difficult than in SVT. It is mainly guided by the cardiac status:

- in patients with a normal left ventricle, i.e. idiopathic VT such as so-called 'verapamil-sensitive' or 'fascicular' VT, as well as in the majority of patients with arrhythmogenic right ventricular disease, the prognosis is good and the risk of ablative procedures is low. Some groups think that ablation represents the best choice in this category of patients, when medical therapy is ineffective. However, in idiopathic VT, the vast majority of patients are well controlled by medical treatment and, in right ventricular disease, most of them have several morphologies of VT originating from different sites, thus complicating the ablation procedure. In that latter case, it should be realized that ablation may represent only one step of the treatment, and not in any way a definitive curative therapy. On the other hand, antitachycardia pacing has been used in idiopathic VT in patients with automatic devices, since the risk of acceleration is very low. Even in that favorable situation, if antitachycardia pacing is indicated, we prefer to use the manual mode, triggered by the physician and not by the patient. Indications of this technique are rare and restricted to a few patients with well-tolerated VT, thus avoiding repetitive hospitalizations.
- in patients with a more or less severely impaired left ventricle, i.e. the majority of VT patients, automatic antitachycardia pacing is prohibited without the rescue of an implanted defibrillator. The manual mode is less useful than in the precedent case, because of the bad tolerance of VT and, often, a high rate of recurrence. Because surgical attempt have a significant mortality risk, some groups have been attracted by ablative techniques. Unfortunately, in the present state of the art, the results of VT ablation are deceptive in these patients. That is why, along with several other groups, we usually prefer to implant an automatic defibrillator, preferably using the antitachycardia pacing mode as a first-line therapy. This therapy is efficient, with a low mortality risk, as will be explained by Dr Mower in the next chapter. It is applicable to the majority of patients: only those with incessant VT or with intractable heart failure should not be considered. In these situations, cardiac transplantation represents the ultimate therapy.

REFERENCES

1. Feld M, Fisher J, Brodman R, Golier F (1987) Coronary sinus rupture complicating catheter ablation of the atrioventricular junction. *J Electrophysiol* 1: 257—260.
2. Bardy GH, Ivey TD, Coltorti FF, Stewart RB, Johnson G, Greene HL (1988) Developments, complications and limitations of catheter-mediated electrical ablation of posterior accessory atrioventricular pathways. *Am J Cardiol* 61: 309—316.
3. Linker NJ, Ward DE, Davies MJ, Camm AJ (1989) Fatal coronary sinus rupture following attempted catheter ablation of an accessory pathway. *J Electrophysiol* 3: 2—6.
4. Chauvin M, Brechenmacher C (1989) A clinical study of the application of endocardial fulguration in the treatment of recurrent atrial flutter. *PACE* 12: 219—224.
5. Saoudi N, Atallah G, Kirkorian G, Touboul P (1990) Catheter ablation of the atrial myocardium in human type I atrial flutter. *Circulation* 81: 762—771.
6. Gillette PC, Wampler DG, Garson A Jr, Zinner A, Ott D, Cooley D (1985) Treatment of atrial automatic tachycardia by ablation procedures. *J Am Coll Cardiol* 6: 405—409.
7. Langberg JJ, Chin MC, Rosenqvist M, et al. (1989) Catheter ablation of the atrioventricular junction with radiofrequency energy. *Circulation* 80: 1527—1535.
8. Borggrefe M, Budde T, Podczeck A, Breithardt G (1987) High frequency alternating current ablation of an accessory pathway in humans. *J Am Coll Cardiol* 10: 576—582.
9. Roman CA, Friday KJ, Wang X, et al. (1989) Ablation of single and multiple accessory pathways with radiofrequency current. (abstract). *Circulation* 80: II—323.
10. Warin JF, Haissaguerre M, Lemetayer P, Guillem JP, Blanchot P (1988) Catheter ablation of accessory pathways with a direct approach. Results in 35 patients. *Circulation* 78: 800—815.
11. Haissaguerre M, Warin JF, Lemetayer P, Saoudi N, Guillem JP, Blanchot P (1989) Closed-chest ablation of retrograde conduction in patients with atrioventricular nodal reentrant tachycardia. *N Engl J Med* 320: 426—433.
12. Warin JF, Haissaguerre M (1989) Fulguration of accessory pathways in any location: report of seventy cases. *PACE* 12: 215—218.
13. Morady F, Scheinmann MM, Kou WH, et al. (1989) Long-term results of catheter ablation of a posteroseptal accessory atrioventricular connection in 48 patients. *Circulation* 79: 1160—1170.
14. Critelli G, Gallagher JJ, Perticone F, Monda F, Scherillo M, Condorelli M (1985) Transvenous catheter ablation of the accessory atrioventricular pathway in the permanent form of junctional reciprocating tachycardia. *Am J Cardiol* 55: 1639—1641.
15. Ward DE, Rowland E, Wainwright RJ, Camm AJ (1989) Electrical ablation of junctional tachycardias showing a long RP interval. *Eur Heart J* 10: 718—724.
16. Cox JL, Gallagher JJ, Cain ME (1985) Experience with 118 consecutive patients undergoing operation for the Wolff–Parkinson–White syndrome. *J Thorac Cardiovasic Surg* 90: 490—501.
17. Iwa T, Mitsui T, Misaki T, Mukai K, Magara T, Kamata E (1986) Radical surgical cure of Wolff-Parkinson-White syndrome: the Kanazawa experience. *J Thorac Cardiovasc Surg* 91: 225—233.
18. Ross DL, Johnson DC, Denniss AR, Cooper MJ, Richards DA, Uther JB (1985) Curative surgery for atrioventricular junctional ("AV nodal") reentrant tachycardia. *J Am Coll Cardiol* 6: 1383—1392.
19. Fujimura O, Guiraudon GM, Yee R, Sharma AD, Klein GJ (1989) Operative therapy of atrioventricular node reentry and results of an anatomically guided procedure. *Am J Cardiol* 64: 1327—1332.
20. Jackman WM, Friday KJ, Fitzgerald DM, Bowman AJ, Yeung-Lai-Wai JA, Lazzara R (1989) Localization of left free-wall and posteroseptal accessory atrioventricular pathways by direct recording of accessory pathway activation. *PACE* 12: 204—214.

21. Fontaine G, Cansell A, Tonet JL, et al. (1988) Techniques and methods for catheter endocardial fulguration. *PACE* 11: 592—602.
22. Sutton R, Kenny RA (1986) The natural history of sick sinus syndrome. *PACE* 9: 1110—1114.
23. Rosenqvist M, Brandt J, Schüller H (1986) Atrial versus ventricular pacing in sinus node disease: a treatment comparison study. *Am Heart J* 111: 292—297.
24. Denjoy I, Leclercq JF, Letouzey JP, Leenhardt A, Coumel P, Slama R (1988) Prevention of paroxysmal atrial fibrillation by atrial pacing (abstract). *Eur Heart J* 9 (Suppl 1): 93
25. Krikler D, Curry P, Buffet J (1976) Dual-demand pacing for reciprocating atrioventricular tachycardia. *Br Med J* 1: 1114—1116.
26. Leclercq JF, Coumel P, Slama R (1979) Stimulateurs intracorporels pour tachycardies rebelles (maladie de l'oreillette exclue). *Arch Mal Coeur* 72: 1279—1285.
27. Spurrell RAJ, Nathan AW, Bexton RS, Hellestrand KJ, Napphoz T, Camm AJ (1982) Implantable automatic scanning pacemaker for termination of supraventricular tachycardia. *Am J Cardiol* 49: 753—760.
28. den Dulk K, Bertholet M, Brugada P, et al. (1984) Clinical experience with implantable devices for control of tachyarrhythmias. *PACE* 7: 548—556.
29. Jordaens L, van Wassenhove E, Clement DL (1986) An implantable antitachycardia pacemaker with back-up pacing and scanning burst mode. *Eur Heart J* 7: 61—66.
30. den Dulk K, Kersschot IE, Brugada P, Wellens HJJ (1986) Is there a universal antitachycardia pacing mode? *Am J Cardiol* 57: 950—955.
31. Nathan AW, Spurrell RAJ, Camm AJ (1984) Steps toward the development of a safe and effective tachycardia terminating pacemaker. *Eur Heart J* 5: 993—1003.
32. Fahraeus T, Lassvik C, Sonnhag C (1984) Tachycardias initiated by automatic antitachycardia pacemakers. *PACE* 7: 1049—1054.
33. Camm AJ, Davies DW, Ward DE (1987) Tachycardia recognition by implantable electronic devices. *PACE* 10: 1175—1190.
34. Pannizzo F, Mercando AD, Fisher JD, Furman S (1988) Automatic methods for detection of tachyarrhythmias by antitachycardia devices. *J Am Coll Cardiol* 11: 308—316.
35. Schmidinger H, Sowton E (1988) Physiological variation in the termination window of re-entry tachycardia studied by non-invasive programmed stimulation. *Eur Heart J* 9: 997—1002.
36. Peters RW, Scheinman MM, Morady F, Jacobson L (1985) Long-term management of recurrent paroxysmal tachycardia by cardiac burst pacing. *PACE* 8: 35—44.
37. Scheinman MM, Evans-Bell T, et al. (1984) Catheter ablation of the atrioventricular junction: a report of the Percutaneous Mapping and Ablation Registry. *Circulation* 70: 1024—1031.
38. Lemery R, Brugada P, Della Bella P, et al. (1989) Predictors of long-term success during closed-chest catheter ablation of the atrioventricular junction. *Eur Heart J* 10: 826—832.
39. Fontaine G, Tonet JL, Frank R, et al. (1986) Traitement des tachycardies ventriculaires rebelles par fulguration endocavitaire associée aux anti-arythmiques. *Arch Mal Coeur* 79: 1152—1159.
40. Morady F, Scheinman MM, Di Carlo LA Jr, et al. (1987) Catheter ablation of ventricular tachycardia with intracardiac shocks: results in 33 patients. *Circulation* 75: 1037—1049.
41. Garan H, Kuchar D, Freeman C, Finkelstein D, Ruskin JN (1988) Early assessment of the effect of map-guided transcatheter intracardiac electric shock on sustained ventricular tachycardia secondary to coronary artery disease. *Am J Cardiol* 61: 1018— 1023.
42. Borgreffe M, Breitjardt G, Podczek A, Rohner D, Budde T, Martinez-Rubio A (1989) Catheter ablation of ventricular tachycardia using defibrillator pulses: electrophysiological findings and long-term results. *Eur Heart J* 10: 591—601.
43. Leclercq JF, Chouty F, Cauchemez B, Leenhardt A, Coumel P, Slama R (1988) Results

of electrical fulguration in arrhythmogenic right ventricular disease. *Am J Cardiol* 62: 220—224.
44. Okamura K, Olshansky B, Henthorn RW, Epstein AE, Plumb VJ, Waldo AL (1987) Demonstration of the presence of slow conduction during sustained ventricular tachycardia in man: use of transient entrainment of the tachycardia. *Circulation* 75: 369—378.
45. Morady F, Frank R, Kou WH, et al. (1988) Identification and catheter ablation of a zone of slow conduction in the reentrant circuit of ventricular tachycardia in humans. *J Am Coll Cardiol* 11: 775—782.
46. Touboul P, Kirkorian G, Atallah G, et al. (1986) Bundle branch reentrant tachycardia treated by electrical ablation of the right bundle branch. *J Am Coll Cardiol* 7: 1404—1409.
47. Tchou P, Jayazeri M, Denker S, Dongas J, Caceres J, Akhtar M (1988) Transcatheter electrical ablation of right bundle branch. A method of treating macroreentrant ventricular tachycardia attributed to bundle branch reentry. *Circulation* 78: 246—257.
48. Fisher JD, Furman S, Kim SG, Matos JA, Waspe LE (1984) DDD/DDT pacemakers in the treatment of ventricular tachycardia. *PACE* 7: 173—187.
49. Roy D, Waxman H, Buxton AE, et al. (1982) Termination of ventricular tachycardia: role of tachycardia cycle length. *Am J Cardiol* 50: 1346—1354.
50. Charos GS, Haffajee CI, Gold RL, Bishop RL, Berkovits BV, Alpert JS (1986) A theoretically and practically more effective method for interruption of ventricular tachycardia: self-adapting autodecremental overdrive pacing. *Circulation* 73: 309—315.

11. Review of implantable defibrillator therapy

MORTON M. MOWER & SEAH NISAM

The automatic implantable cardioverter-defibrillator (AICD) is a battery-powered, implantable device intended to prevent the sudden cardiac death (SCD) syndrome. Fully two-thirds of coronary artery disease mortality occurs by means of this mechanism. It also occurs in heart disease of other etiologies as well. The problem is of epidemic proportions in the developed countries of the world, and claims nearly half a million victims per year in the United States alone, nearly one death per minute [1]. The figures on the Continent are estimated to be similar. All told, there are more victims in any given two-week period from SCD than from the entire acquired immune deficiency epidemic to date.

By definition, SCD is swift, cardiac in nature, and largely unexpected. Management of patients subject to this disorder is fraught with almost unsurmountable difficulties, since most such persons will die within minutes of the onset of symptoms, long before they are able to reach a medical facility. We now know that the pathophysiology usually reflects grave disturbances in cardiac electrical activity, culminating in ventricular fibrillation (VF) and death. It is well-known, even in lay circles, that the only treatment for an acute episode is prompt electrical defibrillation. A surprisingly large proportion of patients in the United States survive the initial episode and reach a hospital. Of those patients, some 20—30% will leave the hospital alive. The relatively good prognosis of recognized cardiac arrest is undoubtedly due to the emphasis in the United States on bystander cardiopulmonary resuscitation (CPR), and on fast response emergency ambulance teams with defibrillators [2].

Nevertheless, a number of lay misconceptions commonly exist regarding the nature of SCD. Often when a person dies suddenly, many people attribute it as being due to a 'massive heart attack'. Within the recent past, for example, the Commissioner of Baseball and the Mayor of Chicago were both described as having such when the evidence really indicated sudden arrhythmic deaths.

Another misconception is that episodes are frequently due to bradyar-

E. Andries, P. Brugada & R. Stroobandt (eds.), How to face 'the faces' of cardiac pacing, 209—214.

rhythmias. In actuality, the typical sequence of events consists of sudden onset of a ventricular tachycardia (VT) or ventricular flutter, degenerating into VF and finally ending in asystole, an entire episode lasting less than 10 minutes. Primary bradyarrhythmias occur only 10—20% of the time. Moreover, this figure is probably an overestimate because asystoles may well be noted when the ambulance crew finally arrives on the scene even though they had not been the primary arrhythmia [3].

During the latter part of the 1970's, many advances were made in identification of patients at high risk of dying suddenly. The large reservoirs of high-risk groups were principally found to be three. First, there are the survivors of previous SCD episodes. When the arrhythmia was not due to a self-limited or correctable cause, such patients have recurrence rates of approximately 30% within one year. Patients with recurrent VT, and those with unexplained syncope likely to be cardiac in origin, are two other groups which also have frequent recurrences, up to 40% within one year, many being fatal.

These findings gave rise to significantly increased interest in aggressive anti-arrhythmic therapy including drugs, anti-tachycardia pacemakers, cardiac surgery, and most recently the AICD. The basic idea behind the development of this device was that it would provide selected high-risk patients with the means of restoring normal heart rhythm within seconds, without the need for specialized medical personnel or additional equipment.

Other treatment alternatives available for high-risk patients are generally not very satisfactory. There are drugs, either given empirically or guided by electrophysiologic study, and surgical ablative therapy. With regard to drugs, it is now abundantly evident that empiric therapy is associated with recurrence rates as high as 31% in 2—3 years, and that control of premature ventricular beats, presumed to be the trigger mechanism for sustained arrhythmias, does not suffice to prevent serious arrhythmia recurrences. Indeed, the recent prematurely terminated cardiac arrhythmia suppression trial (CAST) amply showed that drugs known to be highly effective in eliminating premature ventricular beats not only failed to protect against sudden death, but were actually associated with excess mortality over and above that of the placebo group [4].

With regard to the administration of beta blocking drugs, while numerous studies have shown that they are modestly effective in reducing sudden death, they have to be given to extremely large numbers of patients in order to achieve this statistical effect. Amiodarone has had widespread use but is associated with recurrence rates of 5 to 39%, depending on whether the patient becomes noninducible or not and, moreover, the chances of the patient being able to remain on that drug for a long time are slim. Even when drug therapy is presumably effective, however, there are many additional factors which may affect the ultimate outcome, such as changing substrate, side effects, expense, and patient incompliance.

Electrophysiologic testing provides two major advantages when used to guide drug therapy. It helps in the selection of the proper drug for a given patient, and it identifies the drug-resistant individual who is at extremely high risk of recurrence unless an effective therapy of some kind or another can be found.

Surgical ablation holds promise of arrhythmic cure in some patients, but is associated with considerable perioperative mortality, and the recurrence rates again can be as high as 8 to 46%. The technique is not at all applicable to patients who have VF or multiple foci, or whose VT's are too rapid or unstable to map.

With regard to AICD treatment, it is important to keep in mind that this clinical modality is only some 10 years old. The first patient implantation was performed at The Johns Hopkins Hospital in Baltimore on February 4, 1980 [5] and the device only became widely available for marketing after the Food and Drug Administration (FDA) gave its approval in 1985 [6]. Especially in light of this, the extremely high acceptance rate of this therapy in only a short time can only be described as striking. This is clearly attributable to its marked degree of effectiveness against the condition. Moreover, a major advantage of an AICD is that it is independent of any changing conditions in the underlying substrate.

The usual deployment of leads for the AICD consists of a right ventricular endocardial lead or two myocardial screw-in electrodes for sensing and two defibrillating patches. These latter come in two sizes; the smaller has a surface area of 13.5 cm^2, and the larger of 27 cm^2. Lower thresholds may often be achieved using larger electrodes, although they may also predispose to greater pericardial irritation and to a higher incidence of atrial arrhythmias [7].

Implantation can be through a number of possible surgical approaches. Many units are implanted through a left thoracotomy, especially when patients have had previous chest surgery so as to avoid dissection at a previously operated site. Since a number of patients have other cardiac surgical procedures (e.g. coronary artery bypass grafting, aneurysmectomy with endocardial resection, prosthetic valve replacement, and myectomy) performed concomitantly at the time of implantation, median sternotomy is also frequently done [8]. Sub-xiphoidal [9] and left subcostal techniques similar to pacemaker lead insertion [10], have also been developed and are becoming increasingly used.

Sensing can be selected to consist of rate and morphology, or only the former. An ideal rate setting is lower than the rate of the tachycardia but higher than the patient's own sinus mechanism is able to reach. If the rates of the tachycardia and the supra-ventricular mechanism overlap, the morphology criterion can, at times, be used to differentiate the two entities. The precise indications for the use of dual-sensing essentially deals with the issue of trading greater specificity at the expense of some loss of sensitivity [11].

The most recent AICD unit, which entered clinical testing in September 1988, is a multiprogrammable integrated circuit version whose parameters can be changed at will by the physician. It is programmed with an external hand-held device similar to a pacemaker programmer. In addition to functions such as activation, interrogation, and programming, the device has memory registers for the number of shocks, charge times, and lead impedances. The number of programmable devices implanted will certainly increase remarkably, since the assembly lines have now completely phased out production of nonprogrammable units.

The new unit measures 10.1 by 7.6 by 2.0 cm, and weighs 240 gms. It delivers discharges which can be individually set by the physician to be from 0.1 to 30 J, with the final two shocks if necessary being 30 J. The low energies are designed to terminate VT, and the high energy ones to convert VF. Programmability also allows the selection of any combination of the currently available detection algorithms, thus eliminating the need for having separate models. Delays in the shock delivery are programmable up to 10 seconds to prevent triggering on any nonsustained rhythms.

Clinical trials have also begun with a transvenous defibrillating lead called the Endotak which is implanted similar to conventional pacemaker leads. This obviates the need for thoracotomy. The system also uses a subcutaneous patch which controls the direction of the shock given through this lead [12]. The system appears to be fairly effective so far, and there is some data becoming available indicating that a biphasic pulse waveform may be able to increase this even more to the point where essentially every patient would likely be implanted without thoracotomy.

The FDA has thus far defined only two groups of patients at high risk of SCD in whom this therapy is considered as indicated. These are: (1) patients who have survived at least one episode of cardiac arrest due to ventricular tachyarrhythmia not associated with acute myocardial infarction, and (2) those without previous arrest who have inducible arrhythmias at electrophysiologic testing which cannot be suppressed by conventional antiarrhythmic therapy. On the other hand, other national bodies are becoming somewhat more liberal in their views. For example, similar to what they did in establishing guidelines for pacemaker insertion, the North American Society for Pacing and Electrophysiology (NASPE) has classified a number of additional possible indications for AICD insertion. The categories are Class I (indicated by general agreement), Class II (frequently indicated), and Class III (not generally indicated).

Up through October of 1989, some 8000 patients have received AICD devices implanted in 310 centers worldwide. The therapy has enjoyed accelerating growth, and a number of centers have series of well over a hundred patients each. The patients have generally been male and late middle-aged with poor left ventricular function. Coronary artery disease is the most frequent underlying cardiac substrate. Freedom from sudden death has been found in excess of 98% at one year, and slightly less than 95% at five years [13]. The five-year death rate from all causes combined is only about 21% [14]. Thus, in comparison with other modalities, AICD therapy yields results far in excess of the other competing therapies [15], and well within the range of costs for other modalities of treatment well accepted by society [16].

Even though there are a number of easily recognized 'risk factors' for the SCD syndrome, at the present time, only a small minority of patients from high-risk groups go on to receive an AICD or even get electrophysiologic testing. Most individuals unfortunately continue to receive only empiric therapy if any at all. Reasons for this are complex but may include the general cardiologist (a) being unaware of the magnitude of the sudden cardiac death problem; (b) reluctant to lose his patient to a specialized center; and (c) being unwilling to admit his own therapeutic defeat. Hopefully, however, as electrophysiology grows as a sub-specialty, and with some change in the general medical attitude towards this field, the number of patients who could be presented for study could be as high as several hundred thousand per year. Even by modestly increasing the number of high-risk patients who are referred, a marked effect on the eventual number of patients benefitting from this therapy should easily be seen.

It also appears, moreover, that the implant criteria, as they presently stand, may need to be liberalized. One area which appears to be too restrictive is in the requirement for inducibility. This situation dates back to 1985 when the Health Care Financing Administration (HCFA) adopted guidelines for Medicare reimbursement requiring arrhythmias to be inducible and unable to be suppressed by antiarrhythmics as conditions in order for AICD implantations to be paid for by the Government. Data now indicates that there are numerous noninducible patients with cardiac arrest or recurrent symptomatic VT who are still clearly at risk of SCD. Some of these may have QT interval syndrome (in both overt and concealed forms), mitral valve prolapse, or primary electrical disease, but many patients with coronary artery disease and nonischemic cardiomyopathy are also not inducible, yet remain at markedly high risk. It is evident that in these various patient groups, there are no therapeutic measurements upon which one can base treatment decisions.

The challenges in caring for the high risk types of patients are first to evaluate them adequately, to establish therapeutic goals and a treatment strategy, and then very importantly, to demonstrate that the chosen treatment is efficacious. The evaluation of the patient should certainly define the anatomic in addition to the purely electrical substrates. A typical workup might well include echocardiogram for ejection function, treadmill test followed by catheterization to determine the degree of ischemia, and signal-averaged electrocardiogram for late potentials followed by electrophysiologic study to determine whether the potential anatomic substrates for the re-entrant circuits are, in fact, able to support such rhythms. In this regard, because its specificity and therefore the negative predictive power of signal averaging is so high, the technique may be of particular value in helping to determine which patients will need to go on to be further studied [17].

In the future, as AICD devices become smaller and more suitable for pectoral implantation, nonthoracotomy lead insertions become a reality, and our ability improves so that the clinician will consistently be able to identify the high-risk patient before the initial cardiac arrest, the major direction that device therapy will take is that of 'prophylactic' implantation. It is quite likely that then we will be able to markedly impact the prohibitive loss of life and productivity presently engendered by the SCD syndrome.

REFERENCES

1. American Heart Association (1987) *Heart Facts 1987.* Dallas, Texas.
2. Cobb LA, Baum RS, Alvarez III, H, et al. (1975) Resuscitation from out-of-hospital ventricular fibrillation: 4 years followup. *Circulation* 52: 223–235.
3. Panadis IP, Morganroth J (1985) Initiating events of sudden cardiac death. *Cardiovasc Clin* 15: 81–92.
4. Bigger JT (1990) The events surrounding the removal of encainide and flecanide from the cardiac arrhythmia suppression trial (CAST) and why CAST is continuing with moricizine. *JACC* 15:243–245.
5. Mirowski M, Reid PR, Mower MM, et al. (1980) Termination of malignant ventricular arrhythmias with an implanted automatic defibrillator in human beings. *N Engl J Med* 303: 322–324.
6. Federal Register, November 15, 1985.
7. Mower MM, Reid PR, Watkins L Jr, et al. (1984) Automatic implantable cardioverter-defibrillator: structural characteristics. *PACE* 7: 1331–1337.
8. Watkins L Jr, Mirowski M, Mower MM, et al. (1981) Automatic defibrillation in man: the initial surgical experience. *J Thor Cardiovasc Surg* 82: 492–500.
9. Watkins L Jr, Mirowski M, Mower MM, et al. (1982) Implantation of the automatic defibrillator: the sub-xyphoid approach. *Ann Thorac Surg* 34: 515–520.
10. Laurie GM, Morris GC Jr, Howell JF, et al. (1976) Left subcostal insertion of the sutureless myocardial electrode. *Ann Thorac Surg.* 21: 350.
11. Winkle RA, Bach SM, Echt DS, et al. (1983) The automatic implantable defibrillator: local ventricular bipolar sensing to detect ventricular tachycardia and fibrillation. *Am J Cardiol* 52: 265–270.
12. Bach SM, Barstad J, Harper N, et al. (1989) Initial clinical experience: Endotak TM-implantable transvenous defibrillator system. *JACC* 13: 65A.
13. Mirowski M, Reid PR, Mower MM, et al. (1984) Automatic implantable cardioverter-defibrillator: clinical results. *PACE* 7: 1345–1350.
14. Winkle RA, Mead H, Ruder MA, et al. (1989) Long-term outcome with the automatic implantable cardioverter-defibrillator. *JACC* 13: 1353–1361.
15. Lehmann MH, Steinman RT, Schuger CD, et al. (1988) The automatic implantable cardioverter-defibrillator as antiarrhythmic treatment modality of choice for survivors of cardiac arrest unrelated to acute myocardial infarction. *Am J Cardiol* 62: 803–805.
16. Kuppermann M, Luce BR, McGovern B, et al. (1990) An analysis of the cost effectiveness of the implantable defibrillator. *Circulation* 81: 91–100.
17. Gomes JA, Winters SL, Stewart D, et al. (1987) A new non-invasive index to predict sustained ventricular tachycardia and sudden death in the first year after myocardial infarction. *JACC* 10: 349–357.

12. The implantable defibrillator

ANDRE WALEFFE

Early clinical reports suggested that the automatic implantable cardioverter-defibrillator (AICD) was effective in the prevention of sudden death in patients presumed to be at risk for life-threatening ventricular tachyarrhythmias [1—3]. Subsequent experience supports these impressions [4—11]. There have been no prospective studies, in which therapy is randomized, comparing the AICD to other treatments. Several less rigorous comparisons are published.

The prognosis of patients with repeated ventricular tachycardia or sudden death, resistant to therapy with antiarrhythmic drug is known to be poor; they frequently experience sudden death (expected 2-year risk of sudden death mortality of 30—50%) [12—16]. Survival in such patients treated with the AICD is quite favorable by comparison [17]. Indeed, some have suggested that the implantable defibrillator should be considered as the treatment of choice (instead of the treatment of last resort) for virtually all patients who have been resuscitated from sudden cardiac death.

Recently, Thomas et al. [18] presented data on all 3610 patients who received the AICD between April 1982 and 5 April 1988. The cumulative rate of survival from sudden death in these high-risk patients was 98% at 1 year and 94% at 5 years. Cumulative survival from death from any cause was greater than 70% at 5 years. Winkle et al. [19] have published the largest series from a single center on results with the AICD. In their 270 AICD recipients, the cumulative incidence of sudden death was 1% at 1 year and 4% at 5 years. The cumulative incidence of death from any cause was 26% at 5 years. This level of efficacy is better than has been reported for any other method of therapy.

Fogoros et al. [20] have demonstrated that the AICD is more effective than amiodarone in preventing sudden death in high-risk patients. Between July 1982 and February 1985, cardiac arrest survivors who did not respond to serial drug testing were referred for either AICD (21 patients) or amiodarone treatment (29 patients), depending on the availability of the AICD at the time of evaluation. The cumulative risk for sudden death after 2

E. Andries, P. Brugada & R. Stroobandt (eds.), How to face 'the faces' of cardiac pacing, 215—220.

years of follow-up was 0% for the AICD recipients and 31% for the amiodarone recipients ($p < 0.003$).

Others, however, have sounded a more cautionary tone. Without questioning the ability of the implantable defibrillator to prevent sudden death, Furman [21] asked investigators to undertake a more careful analysis of the impact of preventing sudden death on overall survival in patients receiving this device. Specifically, he speculated that the patients who use their device are the ones most likely to have severe underlying heart disease. Does the prevention of sudden death in these patients substantially prolong their overall survival?

This concern was amplified by Luceri et al. [22] who reported that 14 (28%) of the 50 patients receiving the implantable defibrillator died at a mean of approximately 9 months after implantation. The mean left ventricular ejection fraction in the patients who died was only 23%. More than 90% of the patients who died had been rescued by their device a mean of 4.5 months before their death. Thus, although the implantable defibrillator can be said to have prolonged the life of most of the patients who died, death nonetheless occurred in many patients with severe cardiac dysfunction and tended to occur relatively early.

Tchou et al. [23] reported a more satisfactory experience with the AICD in patients with poor cardiac function. They presented evidence that the prevention of sudden death in patients with severe cardiac dysfunction significantly prolonges overall survival. By performing an analysis of projected survival, they estimated what the survival rate would have been in patients with the implantable defibrillator, had the defibrillator not been present. Projected survival was calculated by assuming that the first appropriate shock received by a patient with the implantable defibrillator would have resulted in death if the defibrillator had not been present. With such an analysis on 70 patients who received the AICD, they found that the device significantly improved survival (compared with projected survival) in patients with an ejection fraction of either > 30% or < 30%. As the investigators noted, however, their data failed to show the expected difference in projected survival between patients with a high or a low ejection fraction. Because patients with poor ventricular function have been noted by most investigators to have a higher risk of sudden death, the failure to see this phenomenon in the study of Tchou et al. suggests either than an occult peculiarity in patient selection may have partially accounted for the favorable outcome in their patients with a low ejection fraction or that the technique used to estimate projected survival without the defibrillator may have been inappropriate.

Fogoros et al. [24] reported a study which strongly supported the conclusions of Tchou et al. [23]. Using a similar analysis in a larger group of patients (119 AICD recipients), their study confirms that the implantable defibrillator significantly prolongs survival through 3 years of follow-up in patients who receive this device for drug-refractory ventricular tachyarrhythmias, whether they have severe or only moderate cardiac dysfunction.

Furthermore, their data suggest that the technique used to estimate projected survival yields reasonable results.

Their patients with a mean left ventricular ejection fraction of only 23 ± 5% had a cumulative 3-year survival rate of 67 ± 12%. This observed rate compares very favorably with the projected 3-year survival rate of only 6 ± 15% ($p < 0.001$). Thus, the prevention of sudden death in their patients with severe cardiac dysfunction seemed to substantially prolong overall survival.

Implicit in the editorial by Furman [21] is the view that patient with good cardiac function may no receive as much benefit from the implantable defibrillator because they are less likely to use their device than are patients with poor ventricular function. In contrast, the study by Fogoros et al. [24] offers strong evidence that patients with good or moderately depressed ventricular function can substantially benefit from an implantable defibrillator. The cumulative 3-year survival rate of patients in their group B, who had a mean left ventricular ejection fraction of 43 ± 9%, was 96 ± 3% compared with a projected survival rate of 46 ± 8% ($p < 0.001$).

INDICATIONS FOR IMPLANTATION OF THE AUTOMATIC IMPLANTED CARDIOVERTER DEFIBRILLATOR (according to the ACC, 1990)

Listed below are proposed indications for the AICD. They are grouped into three categories: *class I*: indications for which there is little or no controversy; *class II*: indications for which uniform agreement among experts does not exist; and *class III*: indications generally not regarded as sufficient to justify implantation of the AICD. It should be noted that improvements in AICD design and results of well-designed clinical trials may alter indications for the AICD in the near future.

Class I:

1. One or more documented episodes of hemodynamically significant ventricular tachycardia (HSVT) or ventricular fibrillation in a patient in whom electrophysiologic testing and Holter monitoring cannot be used to accurately predict efficacy of therapy.
2. One or more documented episodes of HSVT or ventricular fibrillation in a patient in whom a drug was found to be effective or no drug currently available and appropriate was tolerated.
3. Continued inducibility at electrophysiologic study of HSVT or ventricular fibrillation despite the best available drug therapy, or despite surgery/catheter ablation if drug therapy has failed.

Class II

1. One or more documented episodes of HSVT or ventricular fibrillation in a patient in whom drug efficacy testing is possible.
2. Recurrent syncope of undetermined etiology in a patient with HSVT or

ventricular fibrillation induced at electrophysiologic study in whom no effective or no tolerated drug is available or appropriate.

Class III

1. Recurrent syncope of undetermined cause in a patient without inducible tachyarrhythmias.
2. Arrhythmias not due to HSVT or ventricular fibrillation.

COMPLICATIONS WITH AICD

In general, the risk of AICD implantation are regarded as quite acceptable in appropriately selected patients, given the efficacy of the device and the level of illness of the population of patients receiving the device (most of whom have underlying cardiac disease with depressed ventricular function).

In most series, the operative mortality for AICD implantation is between 1 and 3%. The large series from Thomas et al. [18] lists all of the reported complications which occurred in more than 0.5% of 3610 patients receiving the device. These included infected AICD systems (1.1%), misdirected shocks (0.7%), inappropriate shocks in normal sinus rhythm (0.6%), lead dislodgement (0.6%), and lead fracture (0.6%).

THE FUTURE

The majority of the experience reported has been with the Intec/CPI devices because currently available defibrillator technology was largely developed by these manufacturers and the only detailed information regarding long-term follow-up in the published literature concerns these devices. This experience now encompasses 10 years. Other manufacturers entered into the development of cardioverters and defibrillators considerably later because of concerns over the manufacturability of such devices and whether a market existed for them.

Nonthoracotomy implant approaches will likely undergo progressive improvement with improved lead systems and may be given a substantial boost by pulse generators capable of producing biphasic pulses. A nonthoracotomy implant approach, if proven reliable over time, will undoubtedly increase patient and physician acceptance of implantable defibrillator therapy. A less invasive surgical approach will be mandated for prophylactic use of these devices in high-risk populations.

Improvement in pulse generator technology is already available, including bradycardia and antitachycardia pacing capabilities and extreme flexibility in programming, with programmable energy delivery, rate cut-offs and various additional detection algorithms. Electrogram storage and retrieval may facilitate disclosure of the causes of many unexplained shocks, and atrial sensing will likely aid in the discrimination of the multiple rhythms capable of satisfying existing rate criteria and rate criteria variations. Ultimately, sensors of various types will be incorporated into these devices to enable discrimination of hemodynamically stable from unstable ventricular tachyarrhythmias, enabling further requirements in the concept of tailored anti-tachycardia therapy.

REFERENCES

1. Mirowski M, Reid PR, Mower MM, et al. (1980) Termination of malignant ventricular arrhythmias with an implanted automatic defibrillator in human beings. *N Engl J Med* 303: 322—324.
2. Echt DS, Armstrong K, Schmidt P, Oyer PE, Stinson EB, Winkle RA (1985) Clinical experience, complications, and survival in 70 patients with the automatic implantable cardioverter/defibrillator. *Circulation* 71: 289—296.
3. Mirowski M (1985) The automatic implantable cardioverter-defibrillator: an overview. *J Am Coll Cardiol* 6: 461—466.
4. Marchlinski FE, Flores BT, Buxton AE, et al. (1986) The automatic implantable cardioverter-defibrillator: efficacy, complications, and device failures. *Ann Intern Med* 104: 481—488.
5. Thurer RJ, Luceri RM, Bolooki H (1986) Automatic implantable cardioverter-defibrillator: techniques of implantation and results. *Ann Thorac Surg* 42: 143—147.
6. Holt PM, Crick JCP, Sowton E (1987) Experience with an automatic implantable cardioverter defibrillator. Lancet 1: 551—552.
7. Gabry MD, Brodman R, Johnston D, et al. (1987) Automatic implantable cardioverter-defibrillator: patient survival, battery longevity and shock delivery analysis. *J Am Coll Cardiol* 9: 1349—1356.
8. Borbola J, Denes P, Ezri MD, Hauser RG, Serry C, Goldin MD (1988) The automatic implantable cardioverter-defibrillator. Clinical experience, complications, and follow-up in 25 patients. *Arch Intern Med* 148: 70—76.
9. Kelly PA, Cannom DS, Garan H, et al. (1988) The automatic implantable cardioverter-defibrillator: efficacy, complications and survival in patients with malignant ventricular arrhythmias. *J Am Coll Cardiol* 11: 1278—1286.
10. Winkle RA, Mead RH, Ruder MA, et al. (1989) Long-term outcome with the automatic implantable cardioverter-defibrillator. *J Am Coll Cardiol* 13: 1353—1361.
11. Jordaens L, Waleffe A, Derom F, Rodriguez LM, Clement DL, Kulbertus H (1988) Experience with the automatic implantable defibrillator. *Acta Clin Belg* 43: 209—218.
12. Mason JW, Winkle RA (1980) Accuracy of the ventricular tachycardia-induction study for predicting long-term efficacy and inefficacy of antiarrhythmic drugs. *N Engl J Med* 303: 1073—1077.
13. Roy D, Waxman HL, Kienzle MG, Buxton AE, Marchlinski FE, Josephson ME (1983) Clinical characteristics and long-term follow-up in 119 survivors of cardiac arrest: relation to inducibility at electrophysiologic testing. *Am J Cardiol* 52: 969—974.
14. Eldar M, Sauve MJ, Scheinman MM (1987) Electrophysiologic testing and follow-up of patients with aborted sudden death. *J Am Coll Cardiol* 10: 291—298.
15. Wilber DJ, Garan H, Finkelstein D, et al. (1988) Out-of-hospital cardiac arrest. Use of electrophysiologic testing in the prediction of long-term outcome. *N Engl J Med* 318: 19—24.
16. Freedman RA, Swerdlow CD, Soderholm-Difatte V, Mason JW (1988) Prognostic significance of arrhythmia inducibility or noninducibility at initial electrophysiologic study in survivors of cardiac arrest. *Am J Cardiol* 61: 578—582.

17. Lehmann MH, Steinman RT, Schuger CD, Jackson K (1988) The automatic implantable cardioverter defibrillator as antiarrhythmic treatment modality of choice for survivors of cardiac arrest unrelated to acute myocardial infarction. *Am J Cardiol* 62: 803—805.
18. Thomas AC, Moser SA, Smutka ML, et al. (1988) Implantable defibrillation: eight years of clinical experience. *PACE* 11: 2053—2058.
19. Winkle RA, Mead RH, Gandiani VA, et al. (1989) Long-term outcome with the automatic implantable cardioverter-defibrillator. *JACC* 13: 1353—1361.
20 Fogoros RN, Fiedler SB, Elson JJ (1987) The automatic implantable cardioverter-defibrillator in drug-refractory ventricular tachyarrhythmias. *Ann Intern Med* 107: 635—641.
21. Furman S (1989) AICD benefit (editorial). *PACE* 12: 399—400.
22. Luceri RM, Habal SM, Cartellaws A, et al. (1988) Mechanisms of death in patients with the automatic implantable cardioverter-defibrillator. *PACE* 11: 2015—2022.
23. Tchou PJ, Kadri N, Anderson J, et al. (1988) Automatic implantable cardioverter-defibrillator and survival of patients with left ventricular dysfunction and malignant ventricular arrhythmias. *Ann Intern Med* 109: 529—534.
24. Fogoros RN, Elson JJ, Bonnet CA, et al. (1990) Efficacy of the automatic implantable cardioverter-defibrillator in prolonging survival in patients with severe underlying cardiac disease. *JACC* 16: 381—386.

13. Combined automatic implantable cardioverter-defibrillator and permanent pacemaker systems

ROLAND STROOBANDT, ROGER WILLEMS & ALFONS SINNAEVE

INTRODUCTION

The automatic implantable cardioverter-defibrillator (ICD) has become widely accepted as an effective treatment for patients with ventricular fibrillation or hemodynamically unstable ventricular tachycardia, as it has been proven to effectively prevent sudden death in those patients [1]. However, a small number of patients with implanted defibrillators may die suddenly due to bradyarrhythmias either occurring spontaneously or postdefibrillator discharge.

1. NEED FOR BRADYCARDIA SUPPORT PACING

The delivery of a shock may result in the termination of ventricular tachycardia or ventricular fibrillation but may yield sinus bradycardia, sinus arrest, or complete heart block in a significant number of patients. Occasionally, episodes of bradycardia and asystole may be prolonged and may require immediate backup pacing. Some 10 to 30% of the patients who need an ICD, also require cardiac pacing for the management of associated bradyarrhythmias [2, 3].

Ideally the ICD might eliminate the necessity for antiarrhythmic therapy after device implantation. However, in a large series of patients equipped with an ICD [2], up to 69% of the patients received one or more antiarrhythmic drugs to suppress nonsustained tachyarrhythmias that might cause frequent device discharges. In addition, 55% of the patients were given atrioventricular (AV) node blocking drugs to control episodes of atrial fibrillation or flutter. Antiarrhythmic drugs can alter pacing and defibrillation thresholds and tachycardia rates; rendering antitachycardia pacing more difficult and ICD recognition of the tachycardia impossible if the tachycardia rate declines below the cutoff rate of the ICD. Moreover, they may result in the necessity for ventricular pacing as they may induce conduction dis-

E. Andries, P. Brugada & R. Stroobandt (eds.), How to face 'the faces' of cardiac pacing, 221—227.

turbances in patients with normal atrioventricular conduction prior to the implantation of the device.

2. PACING FOR TACHYCARDIA PREVENTION

The clinical role of permanent antibradycardia pacing for the prevention of ventricular arrhythmias is limited. Rare patients presenting with congenital QT prolongation and torsade de pointes, may benefit from this modality as may patients with bradycardia- dependent malignant ventricular arrhythmias. Ventricular pacing at a rate faster than the intrinsic rhythm, but still in the 'physiologic' range, has been shown to decrease spontaneous ventricular ectopy, when used as a primary antiarrhythmic modality [4]. In a small group of patients with life-threatening ventricular arrhythmias, right ventricular pacing in combination with an antiarrhythmic drug regimen proved successful [5]. However, due to the proarrhythmogenic effect of antiarrhythmic drugs and the poor long-term results, pharmacological treatment of ventricular tachyarrhythmias still remains a matter of debate.

3. PACING FOR TERMINATION OF TACHYCARDIA

Antitachycardia pacemakers are useful, as they can terminate a tachycardia effectively and painlessly in some patients. However, initial enthusiasm with antitachycardia pacemakers was tempered by the realization of the dangers and difficulties associated with their use, particularly in the treatment of ventricular tachycardia. As antitachycardia pacing for ventricular tachycardia is associated with an increased risk of acceleration of ventricular tachycardia or the induction of ventricular tachycardia or the induction of ventricular fibrillation, a backup implantable defibrillator is required in patients equipped with an antitachycardia pacemaker.

4. COMBINED USE OF A PERMANENT PACEMAKER AND AN AUTOMATIC IMPLANTABLE CARDIOVERTER DEFIBRILLATOR

Some patients who need an ICD also require cardiac pacing either to treat associated bradyarrhythmias, post-discharge bradycardia, drug-induced conduction disturbances, or to prevent bradycardia-related ventricular arrhythmias. On the other hand, patients equipped with an antitachycardia pacemaker may also need an ICD to treat tachycardia acceleration. Therefore the concomittant use of pacemaker and an ICD is common and may result in several potentially harmful interactions.

4.1. Potential device interactions

1. *ICD oversensing of the pacemaker stimulus resulting in a false ICD discharge*

Oversensing of pacemaker stimuli by the ICD may lead to *double or triple counting of the heart rate* by sensing the atrial or ventricular pacer stimulus, or the evoked ventricular depolarization [2, 6, 7], or both. This may occur if these signals exceed the 150 ms refractory period of the ICD either because of a long programmed AV delay or a local conduction delay between the pacemaker stimulus and the evoked depolarization in the bipolar ICD sensing electrogram. This double or triple counting may exceed the cutoff rate of the ICD and initiate a charging cycle resulting in the delivery of an inappropriate discharge, even though the underlying cardiac rhythm is neither ventricular tachycardia or ventricular fibrillation.

2. *Pacemaker undersensing of ventricular tachyarrhythmias leading to inhibition of ICD discharge*

Pacemakers may fail to sense ventricular tachycardia or ventricular fibrillation and may therefore continue to asynchronously deliver pacing stimuli during these arrhythmias [6]. This may *inhibit the ICD discharge* as the automatic gain control of the defibrillator adjusts for the high amplitude pacemaker artifacts and may be blinded to the lower amplitude fibrillatory waves. This problem may be more frequent with unipolar pacemakers but may also occur when bipolar pacemaker artifacts are large [8].

3. *Transient pacing and sensing problems after ICD discharge*

Pacemaker nonsensing of ventricular complexes and *lack of capture post defibrillation* has been reported to be frequent in both unipolar and bipolar pacemakers [7]. The mean duration of lack of sensing in the units which failed to sense, appropriately, was 9.1 seconds. Lack of capture has been noted to persist for up to 16 [8] and even >56 seconds [6] before normal pacemaker function resumed.

The reasons for this defibrillator-induced pacemaker malfunction are uncertain [7, 9] but may be due to (a) damage to the integrated circuits, (b) catheter displacement, or (c) transient changes in excitability at the electrode-myocardial interface.

The zener diode, present in permanent pacemakers, prevents the current from flowing back to the pacemaker when critical voltages are exceeded by shunting the current through the pacemaker lead. Although it may protect the integrated circuits of the pacemaker from defibrillation energy, it may also cause local myocardial damage secondary to the high current density delivered to the pacemaker lead tissue interface.

4. *Pacemaker reprogramming by ICD discharge*
Pacemaker reprogramming to a backup asynchronous mode and complete turning off of the device has been noted [6] after discharge of an ICD.

5. *Magnet application*
Pacemaker and ICD may be implanted anatomically so close that magnet application affects both devices. Programming the pacemaker can result in an ICD discharge. Therefore, it may be recommended to transiently inactivate the ICD during pacemaker programming.

6. *Pacing above the ICD tachycardia detection rate*
Antitachycardia pacing can trigger the ICD if the pacing rate exceeds the rate cutoff of the defibrillator.

7. *Intersystem electromagnetic interference*
The most likely event may be the sensing of the telemetry signal of the pacemaker by the ICD, resulting in a spurious discharge.

4.2. Recommendations to minimize interactions between pacemakers and implantable defibrillators

The implantation of both a pacemaker and an ICD in a patient is more complex than that of either device alone. Until the more widespread introduction of combined devices, several precautions can be recommended to prevent untoward potentially disastrous interactions between pacemaker and ICD:

- First, unipolar pacing contraindicates ICD implantation. *Bipolar pacing must be used* as bipolar stimulus artifacts are of much lower amplitude than unipolar spikes and are therefore less likely to be sensed by the ICD.
- Each time an ICD is implanted, consideration should be given to the possibility of later pacemaker implantation. *Pacemaker leads and defibrillator rate-sensing leads should be located in separate chambers as far away from each other as possible.* The ICD rate-sensing leads should be placed epicardially on the left lateral ventricle, as far from the right ventricular apex as possible. Therefore, whenever possible, pacemaker implantation should precede ICD implantation.
- When a pacemaker and an ICD are used concomittantly, a rigorous *screening at implant should be performed for double counting and ICD inhibition* by programming the pacemaker in the asynchronous mode with maximal outputs and AV delay (in dual chamber pacemakers).
- At implant testing, pacemaker reprogramming, correctness of sensing and capture *malfunction following ICD discharge* should be evaluated.
- To ensure successful function of both devices and lack of untoward device interaction, *postoperative electrophysiologic testing* should always

be performed before the patient is dismissed from the hospital. Therefore, all different morphologies of the patient's ventricular tachycardias as well as ventricular fibrillation should be induced.

5. FUTURE DEVICES

The next years will witness the introduction of new generation devices capable of tiered electrical therapy. These devices will possess both anti-tachycardia and antibradycardia pacing, low-energy cardioversion and defibrillation capability. Therefore, many of the problems that exist today while both units defibrillator and pacing are separate, will become obsolete when combined devices become available.

5.1. Pacing mode

Many candidates for an ICD have severely impaired cardiac function and receive drugs that may produce chronic bradyarrhythmias. Therefore, they may be in need of AV sequential pacing.

The effects on central hemodynamics of restoring AV synchrony in patients with impaired left ventricular function remains conflicting. It has been suggested that the contribution of the atrial kick to cardiac output declines as the pulmonary wedge pressure increases [10]. However, in a group of patients with severe heart failure (NYHA classes III–IV) atrioventricular sequential pacing was found to produce a higher cardiac index compared to VVI pacing at comparable pacing rates [11]. In patients with preexisting heart failure and a high degree atrioventricular block, improved predicted 5-year cumulative survival rates have been reported for dual-chamber pacing, as compared to VVI pacing [12].

In patients with sinus node disease, dual-chamber pacing remains the mode of choice, as considerable evidence has accumulated that attests its beneficial hemodynamic and antiarrhythmic effects. A recent review of the literature [13], put a very strong case against ventricular pacing in sinus node disease in terms of occurrence of atrial fibrillation, congestive heart failure, retrograde atrioventricular conduction, and thromboembolism. As many of 50% of the patients with sinus node disease show signs of chronotropic incompetence [14]. This means that about 50% of the patients with sick sinus syndrome would benefit from a rate-adaptive pacing system. Moreover, owing to the abnormalities of autonomic tone and endocrine system in heart failure, a large number of patients display chronotropic incompetence [15]. Therefore, new ICD's should have integrated DDD(R) pacing capabilities.

5.2. Future developments

Future devices with multifunctional capabilities to defibrillate and pace the heart will have a common logic control. This will avoid interactions between both pacing and defibrillating functions. The new generation devices will be able to detect post-discharge bradyarrhythmias and to temporarily increase its output. In the future, methods for the automatic measurement of the pacing threshold without loss of capture might provide more safety for the patient while saving energy. New algorithms improving the discrimination between supraventricular and ventricular tachycardias should eliminate inappropriate shocks. Therefore, it may be necessary to elaborate methods to detect the direction of depolarization wavefronts in both the atrium and the ventricle.

The incorporation of a hemodynamic sensor in future devices may distinguish life-threatening arrhythmias from hemodynamic stable ones. This will allow staged electrical therapy and offer the possibility to respond to one tachycardia with one type of therapy.

The use of noncommitted devices, preventing the delivery of a shock when detecting a nonsustained ventricular tachycardia which ends before the capacitor is fully charged, may avoid unnecessary discomfort for the patient. Moreover, recuperation of the energy already stored in the capacitor will definitely increase the longevity of the device, since energy is no longer wasted on a dummy resistance. Further improvements in lead design and search for the most efficacious modality of energy delivery, will certainly result in a more widespread use of nonthoracotomy implants. This will undoubtedly enlarge its application, even to patients with a severely impaired left ventricular function.

REFERENCES

1. Mirowski M (1985) The automatic implantable cardioverter-defibrillator. *J Am Coll Cardiol* 6: 460—466.
2. Winkle RA, Mead RH, Ruder MA, et al. (1989) Long-term outcome with the automatic implantable cardioverter-defibrillator. *J Am Coll Cardiol* 13: 1353—1361.
3. Furman S (1989) Combined automatic implantable cardioverter-defibrillator and pacemaker systems. *J Am Coll Cardiol* 13: 132—133.
4. Fisher JD, Teichman SL, Ferrick A, et al. (1987) Antiarrhythmic effects of VVI pacing at physiologic rates: A crossover controlled evaluation. *PACE* 10: 822—830.
5. Ector H, Van Brabandt H, De Geest H (1984) Treatment of life-threatening ventricular arrhythmias by a combination of antiarrhythmic drugs and right ventricular pacing. *PACE* 7: 622—627.
6. Calkins H, Brinker J, Veltri E, et al. (1990) Clinical interactions between pacemakers and automatic implantable cardioverters-defibrillators. *J Am Coll Cardiol* 16: 666—673.
7. Cohen AI, Wish MH, Fletcher RD, et al. (1988) The use and interaction of permanent pacemakers and the automatic implantable cardioverter defibrillator. *PACE* 11: 704—711.

8. Troup PJ (1986) Lead system selection, implantation, and testing for the automatic implantable cardioverter- defibrillator. *Clin Progr Electrophysiol Pacing* 4: 260—276.
9. Slepian M, Levine JH, Watkins L Jr, et al. (1987) Automatic implantable cardioverter defibrillator/permanent pacemaker interaction: Loss of pacemaker capture following ICD discharge. *PACE* 10: 1194—1197.
10. Greenberg B, Chatterjee K, Parmley WW, et al. (1979) The influence of left ventricular filling pressure on atrial contribution to cardiac output. *Am Heart J* 98: 742—751.
11. Reiter MJ, Hindman MC (1982) Hemodynamic effects of acute atrioventricular sequential pacing in patients with left ventricular dysfunction. *Am J Cardiol* 49: 687—692.
12. Alpert MA, Curtis JJ, Sanfelippo JF, et al. (1986) Comparitive survival after permanent ventricular and dual chamber pacing for patients with chronic high degree atrioventricular block with and without preexistent congestive heart failure. *J Am Coll Cardiol* 7: 925—932.
13. Camm AJ, Katritsis D (1990) Ventricular pacing for sick sinus syndrome: A risky business. *PACE* 13: 695—699.
14. Rosenqvist M (1990) Atrial pacing for sick sinus syndrome. *Clin Cardiol* 13: 43—47.
15. Francis GS, Goldsmith SR, Ziesche S, et al. (1985) Relative attenuation of sympathetic drive in patients with congestive heart failure. *J Am Coll Cardiol* 5: 832—839.

14. Surgical treatment of cardiac arrhythmias. The physician's point of view

PEDRO BRUGADA, FRANCIS WELLENS, PAUL NELLENS, SINAN GÜRSOY, JACOB ATIÉ, GUNTER STEURER, ERIK ANDRIES & HUGO VAN ERMEN

INTRODUCTION

The present therapeutic armentarium for the management of cardiovascular disease is enormous. We have at our disposal many pharmacologic and nonpharmacologic means to diminish or relieve ischemia, improve mechanical function of the heart, and to prevent or terminate an acute episode of cardiac arrhythmia. There are not, however, many cardiovascular diseases which can be truly cured. 'Cure' means the total suppression of symptoms, no need for any additional treatment, return of life expectancy to normal and an improvement in quality of life to such an extent that the patient can really forget that he (she) was ever ill.

Some cardiovascular diseases are truly curable and in the area of cardiac arrhythmias, surgery can cure several of them. Other nonsurgical potentially curative methods are under investigation. Patients with an accessory pathway can be cured [1, 2]. Patients with intranodal reentry tachycardia can also be cured. That is also the case in selected patients with atrial or ventricular tachycardia [3, 4]. The attractiveness of curative solutions, as opposed to palliative treatments, cannot escape the attention of the keen physician, even when curing requires an apparently aggressive procedure.

THE WOLFF–PARKINSON–WHITE AND RELATED SYNDROMES

The first arrhythmia which was successfully operated on was circus movement tachycardia caused by an accessory pathway [5, 6]. The cure of this condition requires a rather simple solution: sectioning the accessory pathway. That prevents any further arrhythmias related to this extra connection. However, this simple solution has required more than 20 years of continuous refinement to arrive to its present stage of development and safety.

Initially, the accessory pathway was approached endocardially (Figure 1).

E. Andries, P. Brugada & R. Stroobandt (eds.), How to face 'the faces' of cardiac pacing, 229–239.

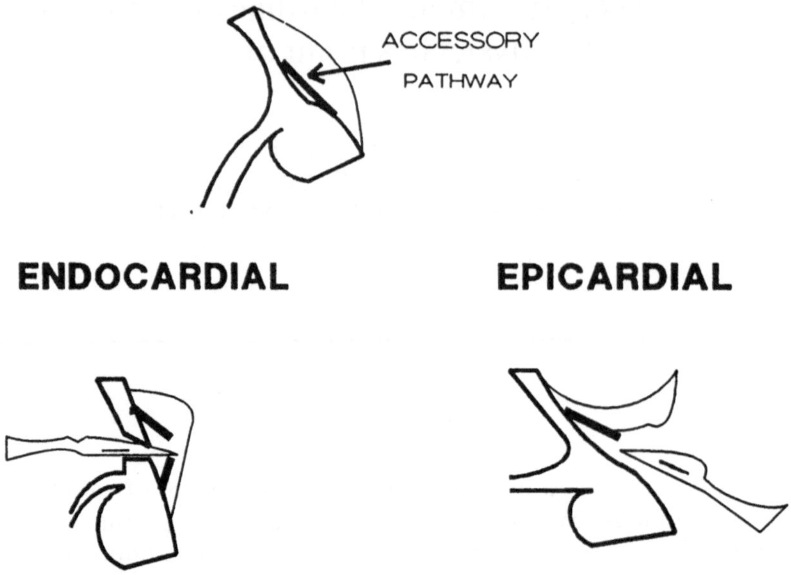

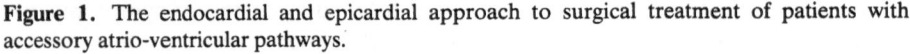

Figure 1. The endocardial and epicardial approach to surgical treatment of patients with accessory atrio-ventricular pathways.

That required cardioplegic arrest and the support of cardio-pulmonary bypass, with all the potential complications involved in clamping the aorta. However, even in the initial series, mortality was very low and generally limited to patients with other associated cardiovascular diseases. Contrarily, the morbidity of the procedure was probably underestimated [7]. The endocardial approach may cause valvular dysfunction, facilitate local endocardial thrombosis and embolism, formation of aneurysms in the area of dissection, and because of muscular damage, cause ventricular tachycardia (unpublished observations). The endocardial approach also has the limitation of dissection done in the arrested heart, without any possibility for continuous electrocardiographic control of the results. Temporary trauma to the accessory pathway may simulate a successful dissection. Additional accessory pathways may be missed after rewarming. That can account for the higher percentage of reoperations reported using the endocardial as compared to the epicardial approach [8, 9].

The epicardial approach [8] (Figure 1) has the advantage of a controlled dissection in the beating heart. The heart is not cooled and the electrocardiogram can be used as a continuous monitoring tool to observe the efficacy of dissection. Additional accessory pathways may become evident immediately after section of the first accessory pathway. The epicardial approach avoids all possible complications associated with cardioplegia, cardiopulmonary bypass, and clamping of the aorta. Damage to the coronary arteries and valvular apparatus are also avoided. Postoperative recovery is certainly faster after epicardial than after endocardial dissection of an accessory pathway.

Given the present state of knowledge, it can be said that the epicardial approach should be the procedure of first choice, followed by endocardial dissection only in the rare cases where the accessory pathway cannot be dissected from the outside. Unfortunately, most surgeons do not master both techniques, particularly those who use the endocardial approach as first choice. There have been extremely negative, unwarranted comments about the epicardial approach by those surgeons who support the endocardial approach. These comments were caused by the lack of knowledge about several important 'tricks' for a successful epicardial approach, including, among others, a skin incision which is a little longer than in a regular thoracotomy so that the posterior part of the heart can be well exposed for the dissection of left-sided and posterior septal accessory pathways. Another necessary 'trick' is to be able to lift the heart without causing severe hypotension. A 'cradle' must be built with a piece of tissue or using a stitch to maintain traction and rotation of the heart. The heart should never be lifted with the hand. Doing so limits diastolic filling and reduces cardiac output (unpublished observations).

Irrespective of these minor differences in possible morbidity, surgical treatment of the Wolff—Parkinson—White syndrome and associated conditions (concealed accessory pathways) can provide a cure for more than 90% of patients, with the current series reporting 0% mortality and 100% success. Reoperation rates are presently below 4%. Our experience in Aalst during the past 6 months is summarized in Table 1 and Figure 2. These operations were performed with the help of Prof. Gérard Guiraudon. A patient with severe pump failure and ischemia who required emergency coronary bypass surgery and had two accessory pathways, could not undergo complete pre or peroperative mapping and so is not included. The endocardial approach was required in three patients: In one with an anteroseptal para-Hisian accessory pathway, in another with a truly endocardial right paraseptal muscular bridge, and in a third patient with a 'Mahaim-like' [10] endocardially located right-free wall accessory pathway with long anterograde conduction times. Apart from this 'Mahaim-like' accessory pathway, one patient had an accessory pathway with long retrograde conduction times ('Coumel' accessory pathway) [11]. The remaining patients all had accessory pathways with fast conduction ('Kent' type). In two patients, skeletonization of the atrio-ventricular node [12] was simultaneously performed because of intranodal reentry.

Forty patients are truly cured and receive no antiarrhythmic drugs. One patient required implantation of a DDD pacemaker for four months because of poor conduction over the normal atrio-ventricular conduction system, but

Table 1. Treatment and followup of 46 patients with an accessory pathway.

Surgery	39 (40 accessory pathways). All cured
Medical	0
Radiofrequency	1
No treatment	6 (Insufficient symptoms)

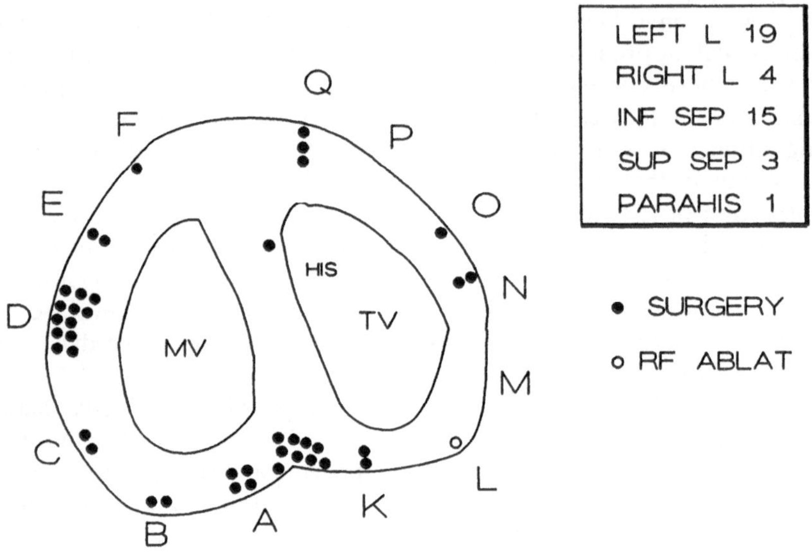

Figure 2. Localization of 42 accessory pathways operated on in Aalst in the past 6 months (one patient had radiofrequency ablation).

was explanted later on. There were two instances of post-operative pericarditis and one patient had transient mitral insufficiency of unclear cause (an endocardial approach had been used in this case). At followup, all patients are asymptomatic and those having reached the 3-month followup have been discharged from further cardiologic control.

The results obtainable with surgery are in sharp contrast with results of medical treatment of patients with accessory pathways. We have previously shown [13] that medical treatment renders completely asymptomatic less than 40% of patients in the long term. The remaining 60% continue to suffer from paroxysmal palpitations in spite of medical treatment, with half still requiring hospital readmission because of long-lasting episodes of tachycardia. Even for those patients in whom arrhythmias do not recur with medical treatment, results cannot be considered a cure. Patients require a daily intake of medication for the rest of their lives. They are exposed to the side-effects of the drugs, and have to be carefully followed-up with cardiologic controls at least every 3 to 4 months. Taking medication daily and having to attend regular controls cannot be equated with curing the disease. Additionally, medical treatment imposes a series of restrictions on sport and other activities, thus limiting the quality of life in these patients.

From the socio-economic point of view, we have also shown that surgical treatment has a much better cost-to-benefit ratio than medical treatment [13]. Twelve years of medical treatment generate the same costs as surgery, and only 40% of patients are asymptomatic. The average patient with the Wolff–Parkinson–White syndrome is 30 years old and life-long medical treatment costs more than 10 times that of surgery. Frequently, however, because of the failure of antiarrhythmic drugs, medical treatment only delays definitive surgical cure, adding unnecessary costs to the total. Given the present state of the art of epicardial surgery for the Wolff–Parkinson–White syndrome, it can be stated that, unless contraindications to surgery exists, the treatment of choice is epicardial dissection of the accessory pathway. The issue of catheter ablation will be discussed later on.

INTRANODAL REENTRY TACHYCARDIA

The first successful operations to cure intranodal reentry tachycardia were reported from Australia [14]. Later on, several variants of the surgical approach were presented, including cryosurgery, simple dissection combined with cryosurgery, and the so-called 'skeletonization' of the atrio-ventricular node [12, 15, 16]. How surgery can cure intranodal reentry tachycardia is unclear. It has been suggested that an atrio-nodal accessory pathway is destroyed at the time of surgery, accounting for the control of reentry [14]. Although this explanation is very exciting, it has to be realized that the atrium has been shown not to be a necessary link in many cases of intranodal reentry. Also, the atrio-ventricular node is not a well-defined structure which can be identified at the time of surgery. At the most, one may guess where the compact part of the node is located. The atrio-nodal approaches cannot be identified. Whatever the mechanism, it is possible to sufficiently modify that structure so as to further prevent reentry while preserving anterograde conduction over the remaining node pathways. The result is control of the arrhythmia without the need for pacemaker implantation, thanks to the preserved anterograde conduction.

With this type of surgery, the only remaining questions arise from a lack of long-term followup studies of the patients operated on so far. Atrio-ventricular block, because of degenerative diseases of the conduction system, is typically a problem of old age. At the present time, it cannot be rejected that surgery of the atrio-ventricular node may increase the incidence of late atrioventricular block. These patients may require permanent pacemaker implantation after 10 or 15 years. If this turns out to be the case, we cannot speak of a real cure of the disease. Only future long-term followup studies will show how many patients are truly cured using this approach. Meanwhile, we believe that one may remain somewhat reluctant and reserve surgery for really uncontrollable cases where an atrial antitachycardia pacemaker is contraindicated for one reason or another. Antitachycardia pacemakers are very effective, can be used alone (without the need for additional medical treatment), and allow us to await for the long-term results of surgery before deciding on operating in very young patients. Table 2 summarizes the results of treatment of intranodal reentry tachycardia in the patients seen in Aalst over the last 9 months. The 7 patients with an antitachycardia pacemaker (Intertach-II, Intermedics Inc.) do not receive any antiarrhythmic drug and none has felt any of the multiple episodes of tachycardia. Only one episode of tachycardia was not terminated because of inactivation of the unit by an electromagnetic field.

ATRIAL TACHYCARDIA

Atrial tachycardia is frequently difficult to treat. That is also true when surgery is considered in medically refractory cases. There is no doubt, however, that directed surgery can cure the disease in the patient with the arrhythmia from a single site of origin. Unfortunately, many patients seem to have many different potential sites of origin of atrial tachycardia, further complicating a direct approach of the problem. It has also to be realized that asymptomatic atrial arrhythmias are a very frequent finding during long-term electrocardiographic monitoring in patients and normals. That makes the assessment of results of surgery very difficult.

Candidates for surgical treatment of atrial tachycardia are symptomatic patients and particularly when they suffer from a single type of atrial arrhythmia which is incessant. This type of arrhythmia does not respond to antiarrhythmic drugs and the persistent high ventricular rate may be responsible, in the long term, for cardiac dilatation and failure ('tachycardiomyopathy'). Excision of the atrial tissue responsible for tachycardia can not only cure the patient from the arrhythmia, but also from the tachycardiomyopathy. Left ventricular function can recover as soon as one month after successful surgical treatment.

ATRIAL FLUTTER AND FIBRILLATION

Isolated atrial flutter is a rare arrhythmia. On the contrary, atrial fibrillation is probably the most common cardiac arrhythmia seen by general cardiologists. The limits between atrial flutter and fibrillation are sometimes difficult

Table 2. Treatment and followup of 38 patients with intranodal tachycardia.

Surgery	4 (All cured)
Medical	26
Antitachycardia pacemaker	7
Other	1

to establish. The two arrhythmias frequently coexist. Treatment in these cases is directed to preserve sinus rhythm (an almost impossible task in the long term), control of ventricular rate (possible in the majority of patients), and prevention of embolic complications (always possible when sinus rhythm is preserved, but unpredictable when atrial fibrillation becomes chronic).

Attempts to surgically cure atrial flutter have met with variable success. Discrete cryosurgical lesions may only change the reentry pathway but fail to prevent a macro-reentry arrhythmia based upon functional properties partly determined by the anatomic anisotropic structure of the atrium [14]. More drastic approaches are required to solve the electric problem.

The 'corridor' operation, one of the many brilliant developments by Guiraudon and coworkers [15], can deal with the problem of atrial flutter and fibrillation in the structurally normal and abnormal heart. After the creation of a corridor between the sinus node and the atrio-ventricular node, normal chronotropy is preserved. Unfortunately, because parts of the atrium may still fibrillate, this operation cannot completely prevent all potential embolic complications.

The 'corridor' operation is a truly life-improving option for patients with uncontrollable atrial flutter or fibrillation. It has several advantages when compared to percutaneous electrical or chemical ablation of the atrio-ventricular conduction system, because restored sinus chronotropism is more physiologic and obviates the need for a pacemaker. In the few patients who required a pacemaker after the 'corridor' operation, a sick sinus syndrome was the most likely cause.

A complete 'cure' for atrial flutter and fibrillation is not yet on the horizon and a lot of research is still required. Meanwhile, the corridor operation can provide sufficient improvement in the quality of life so as to become an option in many patients. Results of surgical treatment of atrial flutter and fibrillation in Aalst are summarized in Table 3.

VENTRICULAR TACHYCARDIA

Surgical treatment can be curative in selected patients with sustained ventricular tachycardia. Ideal candidates are patients with an excellent ventricular function and sustained ventricular tachycardia in whom the arrhythmia can be pinpointed to a discreet ventricular site. Unfortunately,

Table 3. Treatment and followup of 30 patients with atrial arrhythmias.

Surgery	6 (4 'corridor' operations)
Medical	20
Transcoronary ablation	2
Radiofrequency ablation	2

these conditions are met in only a very limited number of patients. The most common etiology of sustained ventricular tachycardia still remains a past myocardial infarction. These patients usually have extensively damaged myocardial areas and not a discrete aneurysm. The larger the area of infarction, the more likely it is that multiple potential reentry circuits will exist. That means the larger the infarction area, the larger the resection required to ensure effective postoperative control of the ventricular arrhythmias. In practice, that means the larger the myocardial infarction, the more likely postoperative complications, including death, will occur. Present-day statistics on the value of surgical treatment of ventricular tachycardia have proven these hypothesis beyond any doubt [16]. Peroperative mortality can be as high as 15% in the most experienced hands. Very low mortality rates are biassed by patient selection and do not reflect the risks involved in surgery for ventricular tachycardia.

In spite of these limitations, there is a role for surgery in the treatment of patients with ventricular tachycardia. The procedure can be curative, provided the appropriate candidates are selected.

Surgical treatment of ventricular tachycardia has also been attempted for etiologies of an arrhythmia different from myocardial infarction [16]. However, variable success rates and high mortality in the case of right ventricular dysplasia, have resulted in the virtual abandonment of surgery for this condition.

Seen from the physician's point of view, there is no doubt that when surgery can offer cure, that should be the preferred approach. However, we need much better selection criteria to recognize candidates for surgery upon whom it can be performed at very little risk and with a high percentage of success.

One surgical option which is frequently forgotten in discussions relating to cardiac arrhythmias, is cardiac transplantation. In patients with recurent cardiac arrest or sustained ventricular tachycardia which are incessant and truly uncontrollable with other therapies, cardiac transplantation may be the only alternative. Cardiac transplant can also be considered in patients with ventricular arrhythmias simply because of cardiac failure. Reality is, however, much more difficult than theory, the major limitation being the availability of donors.

VENTRICULAR FIBRILLATION

At present, no curative treatment exists for recurrent episodes of ventricular fibrillation occurring as a primary arrhythmia. When ventricular fibrillation results from a sustained ventricular tachycardia or from arrhythmias such as atrial fibrillation with fast ventricular rates over an accessory pathway, treatment has to be directed to the cause. However, there are patients who directly suffer from ventricular fibrillation without other preceding abnormal rhythms. The implantable automatic cardioverter-defibrillator (AICD) is an excellent treatment for these patients, but is not curative.

Table 4 summarizes our experience in Aalst in the treatment of ventricular arrhythmias during the past 9 months.

POTENTIALLY CURATIVE NONSURGICAL APPROACHES. ELECTRICAL AND CHEMICAL ABLATION

In an effort to further simplify the curative treatment of cardiac arrhythmias, new percutaneous techniques have been and are still being investigated. They include percutaneous electrical ablation using DC shock or radiofrequency energy and transcoronary chemical ablation of arrhythmogenic areas with ethanol [17]. After several years of investigation, these techniques are beginning to find their place in clinical rhythmology. Some of them will probably be abandoned soon, while others may become extraordinarily successful approaches.

Percutaneous electrical ablation using DC shock has not lived up to the initial expectancies because the results obtained for the treatment of ventricular arrhythmias are less satisfactory than those obtained from surgery. In the treatment of patients with accessory pathways, initial results were encouraging until Warrin et al. reported one sudden death and one aborted sudden

Table 4. Treatment of 71 patients with sustained ventricular tachycardia or ventricular fibrillation.

Etiology	Patients	Medical	Surgery	AICD
Old MI	46	44	4	18
Idiopathic	14	13	—	1
RVD	1	1	—	—
CCM	4	4	—	2
Valvular disease	4	2	2	—
Ischemia	2	1	1	—

Abbreviations: CCM: Congestive cardiomyopathy. MI: myocardial infarction. RVD: Right ventricular dysplasia.

death approximately one week after apparently successful ablation. DC-shock ablation has not offered the expected results in the treatment of atrial flutter. That is why new, less aggressive types of energy have been investigated.

Radiofrequency ablation offers excellent results and complications seem minimal to absent. However, no long-term followup studies are presently available. Also, some details require further clarification. For instance, a group in Münster is giving anticoagulation for six months to all patients after ablation (presented at the European Society of Cardiology Annual Meeting, Stockholm, 1990). The reasons for that are unclear, but may relate to the thrombogenic potential of the lesions created. Anticoagulation may, in itself, result in some complications during followup. The question to be asked about these catheter techniques is What are we trying to avoid or improve? Certainly, a catheter technique, utilized during the first electrophysiologic investigation and providing an immediate cure of the problem is highly desirable. However, that should not be at the cost of any more complications than surgery. In experienced hands and using the epicardial approach, the only advantage of catheter radiofrequency ablation may be avoiding a thoracotomy. Because mortality with surgery is zero, catheter ablation should not result in death. The present results are very encouraging in this sense and it seems that radiofrequency ablation will become the first-choice approach in the management of patients with accessory pathways.

The future of transcoronary chemical ablation is much less clear. We developed this technique and have used it only in patients with incessant ventricular tachycardia not controllable by any other means, including surgery [17]. A study in Alabama which offers transcoronary ablation as first choice to try to cure ventricular tachycardia, will show how valid this approach can be.

CONCLUSIONS

Present-day rhythmology is enjoying a superb era where successful treatment can be found in a high percentage of patients. Unfortunately, only a minority of these patients can be offered total cure for their arrhythmia. With the presently available methods and ongoing research, however, it will probably not take that long before the majority of patients can be offered a definitive curative treatment.

REFERENCES

1. Cobb FR, Blumenschein SD, Sealy WC, Boineau JP, Wagner GS, Wallace AG (1968) Successful surgical interruption of the bundle of Kent in a patient with Wolff–Parkinson–White syndrome. *Circulation* 38: 1018–1029.

2. Sealy WC, Gallagher JJ (1981) Surgical treatment of left free wall accessory pathways of atrioventricular conduction of the kent type. *J Thorac Cardiovasc Surg* 81: 698—706.
3. Iwa T, Ichihashi T, Hashizume Y, Ishida K, Okada R (1985) Successful surgical treatment of left atrial tachycardia. *Am Heart J* 109: 160—162.
4. Ostermeyer J, Breithardt G, Borggrefe M, Godehart E, Seigel L, Bircks W (1982) Surgical treatment of ventricular tachycardiac. *J Thorac Cardiovasc Surg* 83: 865—872.
5. Sealy WC, Gallagher JJ (1980) The surgical approach to the septal area of the heart based on experiences with 45 patients with Kent bundles. *J Thorac Cardiovasc Surg* 79: 542—551.
6. Guiraudon G, Klein GJ, Sharma Ad, Yee R (1987) Use of old and new anatomic, electrophysiologic, and technical knowledge to develop operative approaches to tachycardia, pp. 639—652 in: Brugada P, Wellens HJJ (eds.), *Cardiac Arrhythmias: Where to go from here*? Mount Kisco, New York: Futura Publishing Company.
7. Cheriex EC, Jansen EW, Penn OC, Smeets J, Brugada P, Wellens HJJ (1989) Valve function and wall motion after surgery for the Wolff—Parkinson—White syndrome. *PACE* 12: 685.
8. Guiraudon GM, Klein GJ, Gulamhusein S, et al. (1984) Surgical repair of Wolff—Parkinson—White syndrome: A new closed-heart technique. *Ann Thorac Surg* 37(1): 67—71.
9. Cox JL (1985) The status of surgery for cardiac arrhythmias. *Circulation* 71: 413—417.
10. Atié J, Brugada P, Smeets J, Cruz F, Penn O, Wellens H (1989) Further support of a direct atrioventricular connection in patients with so-called 'Mahaim fibers'. *Eur Heart J* 10: 382.
11. Brugada P, Bär F, Vanagt EJ, Friedman PL, Wellens HJJ (1981) Observations in patients showing AV junctional echoes with a P-R shorter than R-P interval. Distinction between intranodal reentry and reentry using an accessory pathway with a long conduction time. *Am J Cardiol* 48: 611—622.
12. Ross DL, Johnson DC, Denniss AR, Cooper MJ, Richards DA, Uther JB (1985) Curative treatment for atrioventricular junctional ('AV nodal') reentrant tachycardia. *J Am Coll Cardiol* 6: 1383—1392.
13. Lezaun R, Brugada P, Smeets J, et al. (1989) Cost benefit analysis of medical versus surgical treatment of symptomatic patients with accessory atrio-ventricular pathways. *Eur Heart J* 12: 1105—1109.
14. Klein GJ, Guiraudon G, Sharma AD, Milstein S (1986) Demonstration of macroreentry and feasability of operative therapy in the common type of atrial flutter. *Am J Cardiol* 57: 587—591.
15. Guiraudon G, Campbell CS, Jones DL, McLellan DG, MacDonald JL (1985) Combined sino-atrial node atrio-ventricular isolation: A surgical alternative to His bundle ablation in patients with atrial fibrillation. *Circulation* 72: 220.
16. Guiraudon G, Klein GJ, Sharma AD, Yee R (1987) Use of old and new anatomic, electrophysiologic, and technical knowledge to develop operative approaches to tachycardia, pp. 639—652 in: Brugada P, Wellens HJJ (eds.), *Cardiac Arrhythmias: Where to go from here*? Mount Kisco, New York: Futura Publishing Co.
17. Brugada P, de Swart H, Smeets J, Wellens HJJ (1989) Transcoronary chemical ablation of ventricular tachycardia. *Circulation* 79: 475—482.

15. Cost-benefit analysis of arrhythmia technology

HEIN HEIDBÜCHEL & HUGO ECTOR

INTRODUCTION

For a few decades now, physicians have used more than herb-extracts or words of support to treat patients with brady- and tachyarrhythmias, since the assortment of powerful drugs and electronic devices used in the treatment has been continually expanding. However, progress has its price. This text is in the form of a reflection upon what we have to pay for alternative forms of arrhythmia treatment and what success we may expect from them. We will particularly focus on the treatment of tachyarrhythmias. Cost-effectiveness estimates for permanent bradycardia pacemaker implantation, based on the known or assumed efficacy in preventing morbidity or mortality, will not be considered in this text [1, 2].

Although antiarrhythmic drugs have become the most extensively used treatment modality for ventricular tachyarrhythmias, it remains a matter of debate in how far and for which patients they are really beneficial. Therefore, the major focus of medical society on these drugs has been to prove that they were indeed effective. Despite numerous investigational reports, it is still not clear where antiarrhythmic drugs cross the boundary from a life-saving tool to a life-threatening prescription. The results of CAST are a recent and well-recognized example of this fact.

Altogether, the costs of antiarrhythmic drug treatment are considerable. The daily use of the drug itself is not very costly (ranging from about $250 per year for 200 mg/d of metoprolol to about $500 per year for 600 mg/d propafenone; official Belgian prices), but the investigations that are essential for the risk evaluation of a given patient, i.e. the stratification of those who could potentially benefit, make drug therapy as a whole much more expensive. Reported estimates about the cost of hospitalization for therapy selection are sparse, but range from $7,000—$12,000 without electrophysiologic testing to $13,000—$42, 000 with testing [3—5]. The advent of the automatic implantable cardioverter-defibrillator (AICD) has increased the interest in the cost-benefit of tachyarrhythmia treatment because of both the prospects

E. Andries, P. Brugada & R. Stroobandt (eds.), How to face 'the faces' of cardiac pacing, 241—254.

of higher efficiency and of substantially larger costs due to the device itself, the need for thoracotomy at implantation, and the restricted longevity of the AICD. Recently, some authors reported their calculations concerning the cost of defibrillator therapy and the expected benefit from it, compared with conventional antiarrhythmic treatment. In the following paragraphs, we will summarize their conclusions and add some preliminary data from our institution. Since cost-benefit analysis primarily depends on the estimated effectiveness of a given therapy, we will briefly discuss the current evidence that implantable defibrillators can be beneficial in the management of tachyarrhythmias.

AVAILABLE DATA ON THE COST AND COST-EFFECTIVENESS OF AICD THERAPY

Survivors of a cardiac arrest due to ventricular tachycardia or ventricular fibrillation have proven that they are at high risk for the recurrence of lethal arrhythmias. This patient group formed the first FDA-approved indication for implantation of an AICD. Therefore, initial estimates about the cost and cost-benefit of the device were based on the management of cardiac arrest survivors. It is known that patients in whom the induction of the tachycardia by electrophysiologic study can be suppressed tend to have less spontaneous recurrences and a better life expectancy [6—10]. O'Donoghue and coworkers [11] evaluated cost and length of primary hospitalizations for a group of 39 patients with a history of aborted sudden death or life-threatening sustained ventricular tachycardia. They compared the conventional approach of electrophysiologically guided therapy in which cardioverter-defibrillators were reserved for drug-refractory patients (n = 32), with direct device implantation for patients in whom no tachycardia could be induced by a first electrophysiologic study (n = 7). Although AICD placement required a surgical intervention with thoracotomy, early implantations resulted in no longer hospital stays than compared with patients for whom electrophysiologic studies could define an effective drug regimen (12.6 d versus 12.0 d). If, however, no effective antiarrhythmic drug could be found and implantation of the AICD had to follow the serial tests, the cost of total hospital stay was more than doubled (26.3 days on average). The cost of primary hospitalization was lowest if an effective drug could be found ($17,200 ± $9,500). It was $40,400 ± $8,300 for the directly implanted group (with inclusion of the expenses for the device), and was $73,400 ± $17,000 for the patients implanted after failed serial drug testing. The average cost for patients in the serial testing group (AICD or not) was $48,900. Therefore, the authors concluded that early implantation of a cardioverter-defibrillator in all patients with aborted sudden death would result in a shorter hospital stay at comparable or even lower cost compared with conventional serial electrophysiologically guided therapy in which device implantation is reserved for only 'drug-refractory' patients. Indeed, if one roughly calculates the total cost for the first hospitalizations of all 39 patients, i.e. the cost of the most widespread current therapeutic strategy, the charges are about $1,957,000. If all patients would have had an AICD after their first electrophysiologic study regardless its result, total charges would only have been about $1,293,000. However, some remarks have to be made: (1) although such a strategy may be sound from an economical perspective, one can wonder whether the same holds true from a medical viewpoint. The choice of treatment depends on its estimated efficacy and safety (per-operatively and later) for a given set of patient characteristics as inducibility or not, suppression or not, and left ventricular function (which was not mentioned in this study), and not on the fact that it is inexpensive. (2) The fraction of patients for whom suppression of the tachyarrhythmia could be attained by serial drug testing was 37% (12 of 32). Comparable percentages were reported in some other studies examining the success of electrophysiologic testing in identifying effective treatment [8—10, 12, 13]. In other series, however, reported percentages for suppression of inducible arrhythmias were around 60% [5, 7, 14] and in one of the largest published series of cardiac arrest survivors it was 72% [6]. Using the latter ratio to assess the hypothetical total cost of a therapeutic strategy with AICD implantation restricted to noninducibles or nonresponders, would be about $1,340,000 (comparable with $1,293,000 now). The average cost for a patient, who is inducible during an initial electrophysiologic study, would be about $33,000 (versus $48,900 in the presented calculations). (3) The calculations of O'Donoghue include only *first* hospitalizations. Total costs of AICD versus antiarrhythmic drug treatment are also dependent on costs during followup [15].

A more extensive cost-effectiveness analysis considering both initial hospitalization and followup, has been published recently [16]. Kuppermann et al. tried to get an idea of the additional costs and of the additional benefits by use of the defibrillator, compared to conventional medical practice. Only pharmacological therapy was considered because this was the most commonly used treatment modality before the advent of the device. The patient group met the 1985 FDA-guidelines for implantation of the AICD, i.e. survivor of cardiac arrest due to hemodynamically unstable ventricular tachycardia or ventricular fibrillation and persistently inducible at electrophysiologic study. The comparison group was collected from a national registry, retaining patients with diagnosis-codes that corresponded to those of patients who would presumably have received an implantable defibrillator if the device had been available at that time. Control patients thus were historical, not well-defined and certainly not matched. The costs were derived from the US national Medicare data base or local data bases, and a group of cardiologists estimated the frequency of repeat hospitalizations, followup visits, ancillary services and use of (concomitant) antiarrhythmic drugs. Efficacy data were obtained from the medical literature or estimated by the expert panel. This resulted in a 'base case' calculation where the cost-effectiveness was defined as the ratio

$$\frac{\Delta C}{\Delta E} = \frac{\text{total cost AICD} - \text{total costs of drug therapy}}{\text{effectiveness AICD} - \text{effectiveness of drug therapy}}$$

The authors calculated that initial hospitalization costs were on average $55,828 for a defibrillator patient, compared with $17,682 for a drug-treated patient (in 1986 dollars). These data are in good accordance with the previously mentioned figures of O'Donoghue et al. (see Table 1). For the base case patients, the total expected cost with or without an AICD were $121,540 versus $88,900, respectively, for a total estimated life expectancy of 5.1 years versus 3.2 years, respectively. Thus, the cost per life-year saved would be $32,560 ($\Delta C$)/ 1.9 ($\Delta E$) = $17,100. Since the calculations included a whole set of estimated variables (as efficacy data, frequency of rehospitalizations, etc.), many of these were systematically changed in the model: estimated costs per life-year saved varied from $7,020 to $29,604; the extremes depended mainly on the estimated frequency of rehospitalizations in the comparison group, as could be expected. Unfortunately, only a few calculations were made with the simultaneous variation of two parameters, and no calculations are available with multiple simultaneous changes: the range of estimates would probably have been much wider. Kuppermann et al. also extrapolated their model to a 1991 scenario, assuming pulse-generator longevity of approximately 5 years and nonthoracotomy implantation. With these assumptions, the base case scenario yielded a figure of $7,400 (in 1986 dollars), and changing different parameters resulted in a range with maximum costs per life-year saved of $19,640; if the ratio of rehospitalizations would be largely in favour of the AICD, device therapy could even become cost-saving. The authors therefore concluded that the cost-effectiveness of the implantable defibrillator is well within the range of currently accepted life-saving technologies, as treatment of mild hypertension ($11,000 to $72,000), treatment of hypercholesterolemia ($23,000 to $240,000) [17], heart transplantation ($26,900), coronary artery bypass grafting ($44,200), liver transplantation ($51,000) [18], coronary care unit hospitalization for suspected acute myocardial infarction ($33,000 to

Table 1. Average costs of primary hospitalizations for various modes of therapy in patients with a history of sustained ventricular tachycardia or ventricular fibrillation.

Modality	Average cost
Drug treatment without invasive studies [3, 4, 45]	$7,000—$12,000
Drug treatment guided by invasive studies [3, 4, 11, 45]	$13,000—$42,000
Implantable defibrillator treatment [11, 16, 20]	$40,400—$55,828

Data are based on the references as indicated.

$139,000), thrombolysis ($35,000 to $800,000) [19], or peritoneal and hemodialysis ($ ± 58,000). This would indicate that as far as cardiac arrest survivors are concerned, and only pharmacologic treatment is considered as an alternative for AICD, the economical impact of device use seems defensible.

Recently, estimates of the cost-benefit of AICD implantation were made for a European implant centre by the group of Camm at St George's Hospital in London (Anderson M, personal communication). The authors stratified their analysis according to the expected prognosis for cardiac arrest survivors on the basis of the responses during electrophysiological studies, left ventricular function, or both, as reported in Wilber's study of 1988 [6]. If all cardiac arrest survivors would receive an AICD, the cost per life-year saved would be about $100,000. Stratification from low to high risk, using serial drug testing and ejection fraction determinations, resulted in a range for costs per life-year saved from more than $1 million (inducible VT/VF that could be suppressed by AAD; EF ⩾ 30%) to about $35,000 (inducible but refractory to AAD and EF ⩽ 30%). These estimates are well above those of Kuppermann et al., although the estimates of cost for other technologies as heart transplantation and dialysis were in the same range ($24,000 to $32,000 per life-year saved) (see Table 2).

We have some preliminary data on the cost of first hospitalization of cardiac arrest survivors who where evaluated and implanted in an academic hospital in Belgium [20]: for four patients, the costs, covering serial electrophysiologic studies, the AICD device, implantation procedure, and postimplant hospital stay, were $34,500, $38,850, $61,000, and $78,800. The average cost was $53,288. Also, these data are in good agreement with the previously mentioned estimates.

AVAILABLE DATA ON THE EFFECTIVENESS OF AICD THERAPY

It is clear from the preceding paragraphs that the available data about the

Table 2. Cost per life-year saved for different therapeutic interventions.

Therapeutic intervention	Cost per life-year saved
Implantable defibrillator	$17,100—100,000
Mild arterial hypertension (diastolic pressure of 95—105 mmHg)	$23,000—$72,000
Hypercholesterolemia	$23,000—$240,000
Heart transplantation	$26,900
Liver transplantation	$51,000
CCU hospitalization for suspected myocardial infarction	$33,000—$139,000
Thrombolysis	$35,000—$800,000
Peritoneal or hemodialysis	$57,000—$60,000

expenses for conventional or device treatment are sparse, and that this is a fortiori so far the analysis of their relative cost-effectiveness. Moreover, the calculated cost per life-year saved varies considerably (from cost-saving to more than $35,000). Any calculation of cost-effectiveness of a certain therapy (antiarrhythmic drug, VT surgery, AICD) must necessarily make use of its estimated benefit. One has to know just how far morbidity and mortality are affected, compared with the natural history of the disease or compared with any other established form of therapy. In fact, we have no data about the natural history of patients with a previous major tachyarrhythmic event. Because of ethical reasons, the majority of them were treated with antiarrhythmic drugs or surgery. Implantable devices that combine therapeutic algorithms with extensive diagnostic functions can hopefully provide more information on the spontaneous evolution of arrhythmia patients, although it will remain difficult to assess whether a VT recurrence would have been lethal or not. In the absence of data about the natural history, AICD therapy can only be compared with conventional treatment, as was done by Kuppermann's or Camm's groups. There is no doubt that the device can do what it was intended to do: to recognize life-threatening arrhythmias and to stop them. The reported incidence of sudden death in large series of implanted patients is less than 2% over 1 year, less than 8% over 3 years or less than 12% over a 5-year followup [21—16] (see Table 3). This is much lower than in series where no AICD was available: 6 to 17% at 1 year, 21 to 34% at 3 years; 28% at 5 years [6, 7, 9, 10, 14, 27—30]. Even if one considers the fact that this is a rough comparison of actual AICD data with historical control groups, and if one accounts for all the differences between the AICD patient groups and the historical series that may affect outcome, the device seems to be effective in preventing sudden death. For instance, the sudden death rate of VT/VF patients with the best prognosis, those with drug-sensitive arrhythmias during serial drug testing, was found to be 10 to 15% over 3 years when treated with drugs [6, 7]. Based on these results, one could be tempted to consider AICD implantation as the antiarrhythmic treatment modality of choice for patients with a considerable risk of sudden death [31]. But the efficacy of the AICD has to be determined by its ability to prolong life, and not only to prevent sudden death. Since most patients with a previous VT/VF died suddenly (62% in Swerdlow's series, 56% in Wilber's series), it was expected that total mortality would also be beneficially affected by the AICD. The reported total mortality at 1, 3, and 5 years of followup for all AICD patients was reported to be 2—10%, 15—28%, and 21—27%, respectively [21, 22, 24—26, 32]. In the historical series, it was 8—28% at 1 year, 27—50% at 3 years, or 49% at 5 years [6, 7, 9, 10, 14, 28—30]. However, not all groups reported the same positive outcome for AICD patients: in a series of 56 consecutive implants [33, 34], 4%, 11% and 25% of patients died suddenly within 1, 3, and 5 years, and the cumulative mortality from all causes was 12, 26, and 48%. In another series, 7% had died suddenly at 1 year and 40% of the patients had died by all causes after

Table 3. Sudden death and total death in patients with a history of ventricular tachycardia or ventricular fibrillation.[a]

	1 year	3 years	5 years
Sudden death			
Antiarrhythmic drugs	6—17%	21—34%	28%
Implantable defibrillator	<2%	<8%	<12%
Total death			
Antiarrhythmic drugs	8—28%	27—50%	49%
Implantable defibrillator	2—12%	15—40%	21—48%

[a] Data for antiarrhythmic drug and implantable defibrillator series come from references [6, 7, 9, 10, 14, 27—30] and [21—26, 33—35], respectively.

3 years of followup [35]. Thus, the variation of outcome within the defibrillator group is considerable. This is not surprising in the light of the numerous parameters (as inducibility by programmed stimulation, response to antiarrhythmic drugs, left ventricular function, underlying heart disease, previous hemodynamically tolerated VT versus syncope or cardiac arrest) that divide patients with a history of VT/VF into subsets with markedly different prognoses. Baseline differences in clinical characteristics can result in important differences in outcome for distinct patient groups. Remarkably, also the operative death rate in the latter series was 14.3%, in contrast to the much lower values reported by others, ranging from 0.0 to 3.8% (for an overview, see Troup [36]). According to the Belgian AICD registry, 7% of the patients died peri-operatively. How far operative mortality influences the final outcome of AICD patients is far from clear, and not all series include operative or post-operative deaths in their results, which can also contribute to the variability of published data. The same concern has been raised for VT surgery, which may be effective but has an operative mortality from 5 to 15%, reducing its overall effectiveness.

It has been suggested that delivery of a shock by the AICD is, in itself, an independent negative prognostic factor: patients who have received shocks have a 5-year total death rate of 67% compared with 7% for the patients without shocks [34]. The same trend toward higher cardiac and total mortality among patients who have already received a shock, has been noted in other series [36], although sudden death rates were not significantly different in one of them [23] and other authors did not find this correlation [25]. This indicates that patients who have a high risk for sudden death (they already experienced an appropriate shock), are also at highest risk of dying from other causes. It has already pointed out by Fisher that drugs or device-therapy may reduce sudden death rates substantially but that overall survival after 5 years of followup remains about 60% [37]. Most patients with a history of cardiac arrest or recurrent sustained VT have severe underlying heart disease. Therefore, sudden death is not the only possible cause of death: they can also die from heart failure or a new ischemic event, for instance. Thus, preventing sudden death does not necessarily have to lead to a comparable extension of their life-expectancy. This could be predicted from former experience: in Swerdlow's series, 62% of the deaths were sudden, but the single most predictive factor for total survival was the severity of heart failure. It is also known that nonresponders to antiarrhythmic drug testing have lower ejection fractions compared to responders, and have both a higher risk of dying suddenly and dying from other causes. Hence, the prognosis of VT/VF patients is dependent on different factors, many of which seem to be predictive for both sudden death and total (cardiac) death. Patients with the highest sudden death rate can theoretically benefit most from effective anti-sudden-death therapy, but they also have a higher risk of dying from other causes, which reduces the overall benefit of the therapy. Therefore, unless treatment groups are (prospectively) matched and randomized, the balance of these counteracting prognostic factors cannot reliably be assessed.

Fogoros reported in 1987 that he had to leave some of his cardiac arrest survivors without the AICD because its availability was severely curtailed. [38]. All patients had drug-refractory arrhythmias that had caused cardiac arrest or loss of consciousness. Finally, they were empirically treated with amiodarone and only if the defibrillator was available, was it associated to the drug-therapy. The risk for recurrent arrhythmias during a two-year followup period was similar between the 21 patients with an AICD and 29 patients without an AICD. Sudden and total death rates were 31 and 45% in the group treated with the drug alone; in the combination group, the rates were significantly lower, 0 and 5%, respectively. Retrospective analysis could not reveal important differences in the clinical characteristics of both groups to explain the differences in outcome. Although the groups were small and the confidence intervals wide, this retrospective study added important data to the assumed effectiveness of AICD therapy, but it did not unequivocally prove it.

Instead of comparing the outcome of actual patients with historical or retrospectively matched controls, some authors have used the AICD patients as their own control. By assessing the appropriateness of a given shock, they constructed 'projected' survival curves, assuming that the first appropriate shock had aborted a sudden death (eventually after correction for a percentage of tachyarrhythmias that would not have resulted in death without the defibrillator) [26, 32]. Both series included patients with a history of cardiac arrest or hemodynamically unstable ventricular tachycardia. The effective 2-year cumulative death-rate for patients with an ejection fraction of $\leqslant$ 30% was 13 to 24% compared with a projected death rate of 43 or 76%, respectively. When left ventricular function was better, mortality after 2 years was 0 and 1% versus a projected mortality of 50 or 54%. The authors also calculated the survival after a first shock which is another interesting way of assessing the benefit for a particular patient: it was 11 months for patients with an ejection fraction below 30%, and 42 months for the patients with a better ejection fraction [26, 39]. In another series, it was, on average, 14 months for all patients who had received shocks [35]. Although these data suggest a clear benefit in overall survival from the AICD, they tend to overestimate the benefit, even after correction. As a corollary, the reported cost-benefit analysis can be too optimistic.

Thus, although total mortality data from historical comparisons or projected survival estimates seem to indicate a benefit for AICD therapy, the extent of this benefit is variable, likely to be overestimated, and not comparable with former data due to the many different factors determining the outcome. The real benefit can only be determined if the different treatment groups are well matched and therapy is applied randomly. Two prospective randomized studies are currently underway: the Canadian Implanted Defibrillator Study and the Hamburg Cardiac Arrest Trial. The latter study randomizes sudden-death survivors to four alternative therapies: amiodarone, propafenone, metoprolol, or AICD. Preliminary data indicate that the sudden death rate is almost zero in the defibrillator group, clearly beneficial over any drug treatment. However, the effect on total mortality was less prominent, mainly because of the perioperative mortality which was about 5%. The data seem to indicate that defibrillator therapy offers improvement in life expectancy, over metoprolol and amiodarone. Contrary to this, the propafenone arm of the study had already to be stopped because of the significantly higher total mortality than in the other treatment groups. We will have to wait for the final conclusions of this study and of the Canadian study before we have our first confident data on how much defibrillator therapy offers to cardiac arrest patients, compared with more conventional treatment. Only then we will really be able to say what was the cost-effectiveness of our interventions.

Given the difficulties in assessing the cost-benefit for patients who are most prone to lethal tachyarrhythmias — the cardiac arrest survivors — it is even more difficult to make predictions about the impact of AICD therapy on the prognosis for other risk groups who have not yet experienced an episode of cardiac death or hemodynamically compromising ventricular tachyarrhythmia. A number of prospective and randomized studies are planned or have started to address the questions of whether prophylactic use of AICD makes sense in selected patients with (1) nonsustained ventricular tachycardia, (2) undergoing coronary artery bypass grafting, (3) after a myocardial infarction, or (4) before a planned heart transplantation. Especially as a bridge to transplant, the AICD can be very cost-effective due to the good long-term prognosis of transplanted patients and the high rate of sudden death in patients on the waiting list: preliminary data indicate that use of the AICD could even be the most cost-effective in this setting, with an estimated cost per life-year saved of about $14,000, compared with the lowest cost in cardiac arrest survivors estimated to be about $34,000, as mentioned above (Anderson M, personal communication).

The results of all these studies will be very important in determining whether AICD therapy has given benefit to the patient group studied, and what has been the cost of achieving that benefit. However, as we have already mentioned above, the prognosis of an individual patient is determined by a whole set of clinical variables. Prospective studies can only compare a few therapeutic strategies in a well-defined patient group: even then, one needs large numbers of patients in order to detect a difference in prognosis and to be confident about the statistical significance of the observed difference [40] (see Table 4). We have to keep in mind that extrapolation of these results to the daily practice of making choices among many different therapeutic strategies (surgery, ablation, drugs, AICD, transplantation) for a specific patient (with his or her particular set of clinical characteristics) will never be totally predictable. But at least, we will have learned more than we know nowadays.

OTHER ASPECTS OF COST-BENEFIT ANALYSIS AND CONCLUSION

We have discussed the issue of cost-effectiveness in the light of prolonging life. More difficult still is to calculate the cost-effectiveness of defibrillator implantation regarding its effect on the quality of life. Serial drug testing is not only time-consuming and costly, it is also a major psychological obstacle for the patient involved.

Drug treatment or surgery can prevent recurrences of the tachycardia, but sustained and hemodynamically tolerated VT's that do not terminate spontaneously will necessitate emergency hospitalization and probably external DC shocks. On the other hand, the implanted defibrillator will not prevent recurrences, but will treat the tachycardias by automatic discharges. The subjective discomfort, caused by these shocks is variable and unpredictable.

Table 4. Number of replications required for a given probability of obtaining a significant result.[a]

Significance 5%	Power 80%	Power 90%
10 versus 15	680	910
20 versus 25	1090	1460
10 versus 20	195	260
20 versus 30	290	390
10 versus 30	59	79
20 versus 40	80	105

[a] Cochran WG, Cox GM (1957) *Experimental designs*, New York: Wiley & Sons, p. 25.

Moreover, inappropriate shocks can lead to fear and they can necessitate rehospitalization. To reduce the incidence of VT recurrences and/or inappropriate shocks, antiarrhythmic drugs can be associated with an implanted AICD, adding an extra factor that complicates cost-benefit comparisons.

Tachyarrhythmia patients manifest higher degrees of both anger and anxiety compared to normal controls or to other patients. Although the acceptance of the defibrillator 6 months after implant was generally high and implantation of an AICD could reduce the state of anxiety, the state of anger remained unchanged [41]. Some patients resume their normal activities after implantation, but 65% of them reduce their physical activity (mainly out of fear of shocks), and 41% diminish their social interactions [42]. Just how far these changes are attributable to the experience of the ventricular arrhythmia itself or to the implantation of the device remains largely to be elucidated [3]. Keren reported that, in a group of 18, no significant differences in anxiety or depression could be found between patients with AICD or conventional drug treatment, but further studies in larger groups are needed [43].

Most of the patients presenting with cardiac arrest or a sustained ventricular tachycardia have a history of ischemic heart disease and/or heart failure, with hospitalizations for previous myocardial infarction(s) or therapeutic measures as CABG, PTCA, or CPR. Cost-effectiveness calculations have to be evaluated against this background: what are the expenses for implantation of a defibrillator, relative to the whole medical history of these patients? We do not have any idea at present of what, on average, has been spent for these patients before they became eligible for AICD implantation, nor do we know these data for the whole population at risk for a cardiac arrest.

As for many other prophylactic or therapeutic medical interventions, is the prevention of life-threatening tachyarrhythmias without 'benefit' in its pure economical sense. Only a small proportion of cardiac arrest survivors seem to resume work, but few data exist on just how large that proportion is [3, 44]. Medicine, of course, is more than just making people productive: to make people live and to let them enjoy their lives is certainly just as important. Medical society must try to define just how far its technology can achieve these goals and at what price. The available data give us an idea about the order of magnitude, but are far from exact. Ideally, each new device should be implanted in the light of a prospective and randomized trial, so that not only its efficacy for a given indication can be evaluated but also its cost. It is the responsibility of the public and not of medical society to decide which goals are priorities and how much should be invested to obtain them. In making these decisions, the public will ask medical society a lot of pertinent questions and we must be able to answer at least some of them with confidence.

REFERENCES

1. Frye RL, Collins JJ, DeSanctis RW, et al. (1984) Guide-lines for permanent cardiac pacemaker implantation. *J Am Coll Cardiol* 4: 434—442.
2. Galassi A, Dottore E (1989) Cost-effectiveness of today's cardiac pacing. *Cardiologia* 34: 113—116.
3. Krucoff M, Chu F, McCallum D, Perry S (1987) New medical technologies in a cost containment environment: implantable antitachyarrhythmia devices. *PACE* 10: 2—20.
4. Hauser RG (1985) The cost of tachyarrhythmia management. *Arch Mal Coeur* 35—37.
5. Mitchell LB, Duff HJ, Manyari DE, Wyse G (1987) A randomized clinical trial of the noninvasive and invasive approaches to drug therapy of ventricular tachycardia. *N Engl J Med* 317: 1681—1687.
6. Wilber DJ, Garan H, Finkelstein D, et al. (1988) Out-of-hospital cardiac arrest. Use of electrophysiologic testing in the prediction of long-term outcome. *N Engl J Med* 318: 19—24.
7. Swerdlow CD, Winkle RA, Mason JW (1983) Determinants of survival in patients with ventricular tachyarrhythmias. *N Engl J Med* 308: 1436—1442.
8. Morady F, Scheinman MM, Hess DS, Sung RJ, Shen E, Shapiro W (1983) Electrophysiologic testing in the management of out-of-hospital cardiac arrest. *Am J Cardiol* 54: 85—89.
9. Waller TJ, Kay HR, Spielman SR, Kutalek SP, Greenspan AM, Horowitz LN (1987) Reduction in sudden death and total mortality by antiarrhythmic therapy evaluated by electrophysiologic drug testing: criteria of efficacy in patients with sustained ventricular tachyarrhythmia. *J Am Coll Cardiol* 10: 83—89.
10. Eldar M, Sauve MJ, Scheinman MM (1987) Electrophysiologic testing and followup of patients with aborted sudden death. *J Am Coll Cardiol* 10: 291—298.
11. O'Donoghue S, Platia EV, Brooks-Robinson S, Mispireta L (1990) Automatic implantable cardioverter-defibrillator: is early implantation cost-effective? *J Am Coll Cardiol* 16: 1258—1263.
12. Wellens HJJ, Brugada P, Stevenson WG (1985) Programmed electrical stimulation of the heart in patients with life-threatening ventricular arrhythmias: what is the significance of induced arrhythmias and what is the correct stimulation protocol? *Circulation* 72: 1—7.
13. Bonnet CA, Elson JJ, Fogoros RN (1990) Prognostic significance of noninducibility during baseline electrophysiologic study versus noninducibility during serial drug testing in cardiac arrest survivors. *J Am Coll Cardiol* 15: 124A.
14. Ruskin JN, DiMarco JP, Garan H (1980) Out-of-hospital cardiac arrest: electrophysiologic observations and selection of long-term antiarrhythmic therapy. *N Engl J Med* 303: 607—613.
15. Winkle RA (1990) Early automatic implantable cardioverter-defibrillator implantation: medical and economic considerations and inequities in health care reimbursement (editorial). *J Am Coll Cardiol* 16: 1264—1266.
16. Kuppermann M, Luce BR, McGovern B, Podrid PJ, Bigger T, Ruskin JN (1990) An analysis of the cost-effectiveness of the implantable defibrillator. *Circulation* 81: 91—100.
17. Martens LL, Rutten FFH, Erkelens DW, Ascoop, CAPL (1990) Clinical benefits and cost-effectiveness of lowering serum cholesterol levels: the case of simvastatin and cholestyramine in The Netherlands. *Am J Cardiol* 65: 27F—32F.
18. Kankaanpää J (1987) Cost-effectiveness of liver transplantation. *Transpl Proc* 5: 3864—3866.
19. Goldman L (1990) Cost-effectiveness perspectives in coronary heart disease. *Am Heart J* 119: 733—740.
20. Patient registry of the University Hospital of the KU Leuven, Leuven, Belgium.
21. Thomas AC, Moser SA, Smutka ML, Wilson PA (1988) Implantable defibrillation: eight years of clinical experience. *PACE* 11: 2053—2058.

22. Winkle RA, Mead RH, Gaudiani VA, et al. (1989) Long-term outcome with the automatic implantable cardioverter-defibrillator. *J Am Coll Cardiol* 13: 1353—1361.
23. Veltri EP (1989) Letter to the editor. *PACE* 12: 1964—1967.
24. Nisam S, Mower M, Moser S (1991) ICD clinical update: first decade, initial 10,000 patients. *PACE* 14: 255—262.
25. Fogoros RN, Elson JJ, Bonnet CA (1989) Actuarial incidence and pattern of occurrence of shocks following implantation of the automatic implantable cardioverter defibrillator. *PACE* 12: 1465—1473.
26. Fogoros RN, Elson JJ, Bonnet CA, Fiedler SB, Burkholder JA (1990) Efficacy of the automatic implantable cardioverter-defibrillator in prolonging survival in patients with severe underlying cardiac disease. *J Am Coll Cardiol* 16: 381—386.
27. Swerdlow CD, Freedman RA, Peterson J, Clay D (1986) Determinants of prognosis in ventricular tachyarrhythmia patients without induced sustained arrhythmias. *Am Heart J* 111: 433—438.
28. Horowitz LN, Josephson ME, Farshidi A, Spielman SR, Michelson EL, Greenspan AM (1978) Recurrent sustained ventricular tachycardia. 3. Role of the electrophysiologic study in selection of antiarrhythmic regimens. *Circulation* 58: 986—997.
29. Myerburg RJ, Kessler KM, Estes D, et al. (1984) Long-term survival after preshospital cardiac arrest: analysis of outcome during an 8 year study. *Circulation* 70: 538—546.
30. Graboys TB, Lown B, Podrid PJ, DeSilva R (1983) Long-term survival of patients with malignant ventricular arrhythmia treated with antiarrhythmic drugs. *Am J Cardiol* 50: 437—443.
31. Lehmann MH, Steinman RT, Schuger CD, Jackson K (1988) The automatic implantable cardioverter defibrillator as antiarrhythmic treatment modality of choice for survivors of cardiac arrest unrelated to acute myocardial infarction. *Am J Cardiol* 62: 803—805.
32. Tchou PJ, Kadri N, Anderson J, Caceres JA, Jazayeri M, Akhtar M (1988) Automatic implantable cardioverter defibrillators and survival of patients with left ventricular dysfunction and malignant ventricular arrhythmias. *Ann Int Med* 109: 529—534.
33. Gross J, Zilo P, Ferrick K, Fisher JD, Furman S (1991) Sudden death mortality in implantable cardioverter defibrillator patients. *PACE* 14: 250—254.
34. Zilo P, Gross JN, Benedek M, Fisher JD, Furman S (1991) Sudden death mortality in implantable cardioverter defibrillator patients. *PACE* 14: 273—279.
35. Myerburg RJ, Luceri RM, Thurer R, et al. (1989) Time to first shock and clinical outcome in patients receiving an automatic implantable cardioverter-defibrillator. *J Am Coll Cardiol* 14: 508—514.
36. Troup PJ (1991) Programmed stimulation-guided therapy compared with implantable cardioverter defibrillator device therapy in the treatment of ventricular tachyarrhythmias. *PACE* 14: 267—272.
37. Fisher JD, Brodman RF, Kim SG, Ferrick KJ, Roth JA (1990) VT/VF: 60/60 protection (editorial) *PACE* 13: 218—221.
38. Fogoros RN, Fiedler SB, Elson JJ (1987) The automatic implantable cardioverter-defibrillator in drug-refractory ventricular tachyarrhythmias. *Ann Int Med* 107: 635—641.
39. Furman S (1989) AICD benefit (editorial). *PACE* 12: 399—400.
40. Cochran WG, Cox GM (1987) *Experimental Designs*, New York: Wiley & Sons. p. 25.
41. Vlay SC, Olson LC, Friccione GL, Friedman R (1989) Anxiety and anger in patients with ventricular tachyarrhythmias. Responses after automatic internal cardioverter defibrillator implantation. *PACE* 12: 366—373.
42. Cooper DK, Luceri RM, Thurer RJ, Myerburg RJ (1986) The impact of the automatic implantable cardioverter defibrillator on quality of life. *Clin Prog Electrophysiol Pacing* 4: 306—309.
43. Keren R, Aarons D, Veltri EP (1991) Anxiety and depression in patients with life-threatening ventricular arrhythmias: impact of the implantable cardioverter-defibrillator. *PACE* 14: 181—187.

44. Vlay SC (1986) The automatic internal cardioverter-defibrillator: comprehensive clinical followup, economic and social impact. The Stony Brook experience (editorial). *Am Heart J* 112: 189–194.
45. Mitchell LB, Wyse DG, Eliasoph HP, Koskosi CD, Duff HJ (1987) A randomized cost comparison on the invasive and noninvasive approaches to drug selection for ventricular tachyarrhythmias. *Circulation* 76 (Suppl IV): 509.

16. Arrhythmia technology. The insurer's point of view

R. VAN DEN OEVER

INTRODUCTION

Advances in science and technology have led to an explosive growth in diagnostic, therapeutic, preventive, and rehabilitative methodologies in health care during the second half of the twentieth century.

Any device, procedure, or pharmaceutical that showed a net beneficial effect without too many side effects or risks, was automatically added, as a new weapon against disease, to the armoury of the health professional who, together with his patient, has always been the first assessor of a technology's value.

The number of new and complex technologies that have become available is troubling the doctor's choice, since new technology rapidly becomes obsolete and is often in competition with older techniques. Where is the health professional to find clear-cut guidelines permitting him to make a correct judgement about which technology to use, which to discard, or which to replace by the latest and often most expensive one?

Although it is generally accepted that modern medical technology has improved both life expectancy and quality of life for mankind, many unanswered economic, social, political, legal, and ethical questions about the appropriate use and distribution face us.

Until recently, clinicians have played a decisive role in the organization, delivery and even financing of health care and, in their controlling function, the health professionals did not pay special attention to policy-making and organization of the health-care delivery system.

Since health technology is believed to have a major impact on health-care cost and most countries are confronted with economical difficulties, there is need to readjust the scarce resources allocated to health care. The former policy, according to which social security should offer a comprehensive coverage regardless of the costs, has been abandoned and the choice between technologies and what society can afford has become concrete and difficult.

The government and social insurers concerned about the increasing cost

E. Andries, P. Brugada & R. Stroobandt (eds.), How to face 'the faces' of cardiac pacing, 255–265.

of medical care and its affection by health technology, worked out cost containment strategies and tried to stimulate and broaden the evaluation process on safety, effectiveness, and the legal and ethical aspects of existing and new technolgoy.

It is striking that identical budgetary problems exist for health care financed by a fee-for-service system as well as for the closed-budget system and, in the contentious debate over uncontrollable increase of costs, all the different elements and partners in the health-care delivery system have each, in turn, been held responsible.

MEDICAL TECHNOLOGY AND COST, FACTS AND FALLACIES

The term 'medical technology' indicates all means of prevention, diagnosis, treatment and rehabilitation including equipment, material, pharmaceuticals, as well as the administration and organization of these activities.

Major factors contributing to growing health expenditure, other than medical technology, are the number of hospital beds, the teaching hospitals, the growing manpower in health care, the consumer's behaviour, and the demographic change.

Policy-makers and insurers have concentrated their different cost-containment strategies successively on some of these issues in order to control the volume of health services offered, the demand for these services, or their prices. Regulation of the number of health workers, physician's fees, or hospital admission costs, turned out to be unsuccessful, or at least insufficient, and so the efforts of cost containment focused on medical technology.

Erroneously, there is still belief that controlling the cost of medical care can be achieved by harnessing the development, the diffusion, and use of the relatively few but rather costly technologies such as organ transplantation, cardiac surgery, and MR imagers.

Even with a high cost per procedure, this big ticket technology represents only a small fraction of the health-care budget and most cost-utility studies favour this medical technology applied to a limited number of patients and with an acceptable cost per quality adjusted year of life gained.

Contrary to this, the more routinely applied small-ticket technology such as laboratory tests in clinical chemistry, conventional X-ray, endoscopies, and other simple diagnostic and therapeutic procedures, weigh heavier on the health budget and is performed by a large group of health professionals for the majority of the population.

This overuse of small-ticket technology is caused by the relative arrears of basic medical science to the evolution of technological knowledge, by the marketing policy of the industry, the growing number of health-care professionals, and the more generalized and direct information by mass media and advertising campaigns creating high expectations and demands on the part of the patient, who does not understand the cost of such technologies and their often limited value in improving the quality of life.

Basic medical science falls behind the technological sciences, which are expressed in the varied and increasing offers in diagnostic possibilities. At an ever-growing speed, new apparatus, techniques, kits, and devices are offered to the medical world, where a multitude of mostly diagnostic data on the patient are produced without essential contribution to the therapy. On the other hand, the medical technology producing industry leads an agressive marketing policy among the health-care professionals, where sales figures prevail over therapeutic necessity and social benefit.

As manpower in health care is growing as a result of too many training centers, a lack of planning, and a superspecialization, it reduces the number of potential patients per provider, and the physician will try to meet his set standard income in forcing up the number of technical procedures paid for by a fee-for-service system.

Financing and investment policy of the hospital and elements of prestige towards the customer also have their influence on inappropriate use of — especially diagnostic — technology, and even medical audit, liability suits, patient's criticism, and second opinions provoke a defensive medicine with a tendency to create a climate of false diagnostic or therapeutic security by a maximal number of tests and examinations.

Finally, the patient plays a role in overconsumption of technical procedures by his increasing demands and expectations towards health care, modern medical science, and total insurance cover. How can any patient still believe that a normal pregnancy can be managed by a family physician without sophisticated technology, when it is a rule to have a private obstetrician, prenatal diagnostic investigations, and monthly echographies? A similar example of the influence of direct demands by the patient is found in the relatively high prescription and consumption rate of pharmaceuticals.

OTHER CONSEQUENCES OF MEDICAL TECHNOLOGY

Beside the influence on an explosive cost evolution because of the high prices paid for purchasing apparatus, highly qualified personnel and the number of procedures, overuse or inadequate use of such routine medical technology as clinical chemistry, ECG, ultrasound or preoperative chest VX-ray has to be reduced by more appropriate information, training, and financing.

Since more health-care professionals are using high technology in medical practise without necessarily having received the proper training at university, questions have to be asked about quality, opportunity, and postgraduate retraining.

The domain of every medical speciality is less clearly defined as diagnostic and therapeutic imaging or endoscopic techniques are becoming the new tools of any physician. Further, a theoretically cost-saving effect of new technology can, in practise, turn out to be the opposite because of the large number of recidives, repetitions, combination of treatments, larger population of sometimes asymptomatic patients, and exorbitant prices of the employed disposables, even if the intrinsic value and efficacy of the technique or device leading to a quicker and more precise diagnosis and treatment, and reduced or even prevented hospital admission cannot be denied.

The rational application of any technology asks for an organized and uniform assessment that can guide a correct choice, a meaningful use according to the best indications and an equitable reimbursement.

Finally, the classical private relationship between doctor and patient has changed and technology has brought the medical team, the technician, and the biomedical engineer into the functioning routine scheme of the operating room and the imaging or radiotherapy department. It is sometimes difficult for the doctor to perceive the patient as a human individual in the often impersonal and cool technological approach focussing on separate organs or body functions.

ARRHYTHMIA TECHNOLOGY ASSESSMENT

In the diagnosis and treatment of cardiac arrhythmias, the development of medical technologies and their effect on outcome, cost, and quality hold a particular place.

In the detection and treatment of the more prevalent arrhythmias, the reimbursement policy of Belgian social security proves that the insurer in this country does not decide for or against a technology prior to its release on the health-care market, thus facilitating assessment to the benefit of all patients. In 1989, a total of 5,244 pacemakers were implanted in a population of 9.8 million. Including the 5% devices that are implanted under warranty, after refurbishing or in patients not covered by health insurance, the number of implants in Belgium will exceed 6,000 in 1990. There were only 10.9% replacements registered in 1989, a considerable decrease compared to the 18% pulse generator change in 1981 (Table 1).

The fact that cumulative pacemaker replacement is not occurring in a higher percentage of all implantations, can only partially be explained by better longevity, but it especially shows major changes in the indications associated with permanent cardiac pacing during the past decade and a still growing number of treatment centres. The sick sinus syndrome, cerebral dysfunction, and heart failure are more frequently diagnosed. Since these conditions are prevalent in elderly patients and pacing has not clearly been shown to influence survival in the sick sinus syndrome, it is essential to critically evaluate the indications for pacing and to accept only high-risk patients [1].

The actual pulse generator provides many options and its costs grow with

Table 1. Pacemaker implantation in Belgium (Source: RIZIV).

	First implant	Replacement	Total
1981	2,349	522	2,871
1985	4,276	527	4,803
1989	4,668	576	5,244

the complexity of the device, so that, for a given condition, careful selection of the appropriate pacing device becomes crucial.

Cardiac pacing for AV-block holds the first position in the cost-effectivity ratings of medical technology. In the cost per quality, adjusted life-year gained, actually estimated at BF 40,000, the pacemaker equals the total hip replacement.

It is obvious that averted morbidity costs and relieved symptoms by permanent cardiac pacing are not quantified, but are considered by health insurance to be of major importance, since reemployment after pacemaker implantation as an element of cost-benefit analysis is irrelevant.

Among 1798 patients receiving a pulse generator, only 236 were aged less than 65 for men or 60 for women and no more than 55 (3%) resumed professional activities. Of these 55 reemployed, 21 received an implant for bradyarrhythmia as the original cause of their unfitness to work and in 34 we found the device only as an accessory therapy.

On the other hand, patient longevity in the youngest age groups requiring pacemakers for surgically induced heart block, sick sinus syndrome, congenital complete heart block, and tachydysrhythmia has importantly improved due to heart stimulation [2]. Since frequent reoperations in this age population are to be expected for replacement of the generator, some authors recommend a renewed interest in nuclear pacemakers that are considered safe and reliable and have their greater initial cost balanced by their longevity and fewer interventions [3].

Because cardiac arrhythmias represent a substantial component of the yearly hospital admissions in Belgium, it is important to evaluate the various procedures and technologies that are applied in the diagnosis and treatment of these conditions.

Sudden death claims an estimated 12,000 lives per year in this country and when death occurs within 1 hour of the onset of the symptoms, 90% are the result of ventricular tachyarrhythmias. The majority of victims are middle-aged men with coronary heart disease, but in approximately 25%, SCD is the first manifestation of the underlying disease.

The detection of premature ventricular complexes by ambulatory monitoring is predictive of an increased risk and so we have today two diagnostic instruments used for identification of the patient at risk: Holter monitoring and electrophysiology (EP).

Total reimbursement for cardiology procedures in 1989 amount to BF 4,212 million of which Holter ECG takes 11% with 131,394 tests (Table 2). Because the arrhythmia may not occur during the monitoring period, the Holter technique is not proved to be completely protective against an established arrhythmia. The fact that Holter can be applied easily and on an outpatient basis with a favourable reimbursement rate, explains its success and extended use by cardiologists.

In the patient population who have survived a life-threatening arrhythmia unrelated to an acute myocardial infarction and who face a recurrence risk of 25% in the next year, Holter monitoring as a therapy guide is limited by false negative responses and the lack of frequent spontaneous arrhythmias. EP studies represent the only way to prognosticate recurrence, to gain insight into the mechanism of arrhythmia, and to guide therapy. However, the number of 1,000 EP tests performed yearly remained rather constant and small between 1985 and 1990 (Table 3). This amazing phenomenon could be explained by the restricted availability of cardiologists adequately trained to do EP studies, the lack of appropriate referrals by insufficiently informed family physicians, and the discouraging reimbursement rate.

A more adequate reimbursement of this valuable diagnostic tool is actually considered by health policy-makers as EP study might be cost-effective by reducing the length of hospitalization and in selecting the appropriate therapy. Recently, outpatient EP testing was shown to be feasible and safe for patients without life-threatening arrhythmia with substantial cost savings being achieved in the better preselection of arrhythmias evaluated in or outside the hospital [4].

Today's treatment for ventricular arrhythmia is based on three forms of medical technology: drugs, surgery, and implantable devices.

Drug treatment is still considered to be the treatment of choice where acute oral drug testing offers a pragmatic approach to the rapid selection of the drug with the best efficacy/side effect ratio in comparing multiple anti-arrhythmic products within the same patient.

Table 2. Principal components of the 1989 expenditures in millions of US$ by health insurance for cardiology procedures (Source: RIZIV).

	US$	*n*
Routine ECG	44.46 (36.42)	2,601,257 (56.52)
Echocardio	26.17 (21.44)	429,940 (9.06)
Holter ECG	13.47 (11.04)	131,394 (2.76)
Vectocardiogram	7.10 (5.82)	356,438 (7.51)
Effort ECG	6.89 (5.52)	273,553 (5.77)
Phonocardio	5.21 (4.27)	211,743 (4.46)

Total reimbursement amount 1989: 122.

Table 3. Total reimbursement for EP investigation (procedure 476092, RIZIV).

Year	US$ mill.	*n*
1985	0.20	840
1986	0.23	914
1987	0.31	1,236
1988	0.26	1,024
1989	0.27	1,030
1990	0.26	998

However, antiarrhythmics present us with numerous questions. Firstly, there are the considerable short-term and long-term side effects such as seizures, hypotension, cardiac depression, corneal deposits and, ironically enough, even the induction of ventricular tachycardia. Secondly, initiating or revising drug therapy requires prolonged hospitalization, repeated EP studies, and is not without risk for the patient.

Finally, antiarrhythmic drug therapy proves to be difficult in the determination of precise therapeutic dose, can have considerable changes of effect over time within the same patient, and fails to manage ventricular arrhythmias in over 25% of the cases.

Total sales figures for antiarrhythmics increase continuously and while the Belgian market is now dominated by only 3 products without a relevant change in drug consumption quantities, only climbing product price is responsible for this (Table 4). Yearly intramural antiarrhythmics consumption does account for not more than BF 20 million.

The interventional approach of ventricular arrhythmia can consist of a resection of the deficient area of the heart or elimination of aberrant conduction by surgery or catheter ablation. However, catheter ablation is still considered experimental and surgery can provide effective control of patients with an accessory atrioventricular pathway leading to symptomatic arrhythmias only when medical treatment fails. A recent study concludes that surgical treatment has a better cost-benefit ratio than drug therapy (total mean cost per successfully treated patient $US11,250 vs $US6,949 after 56 months followup), mainly because a higher number of medically treated patients remain symptomatic [5].

Although nonpharmacologic treatment of arrhythmias carries with it a one-time period of higher peroperative risk, it can be curative and is then preferable to the uncertainty and possibly higher cumulative risk associated with drugs because of inefficacy, noncompliance, or limiting side effects [6].

Because of the still rather few indications for surgery or catheter ablation, we have to wait for substantial improvements in these new technologies before interventional treatment can be considered the first choice in the

Table 4. Sales figures for antiarrhythmic drugs in Belgium for ambulant patients.[a]

Year	Total[a]	Flecainide (%)	Disopyramide (%)	Propafenone (%)	Kinidine (%)
1984	8.69	—	45	—	15
1985	10.14	30	40	—	20
1986	12.17	30	30	—	20
1987	13.04	35	30	15	—
1988	13.91	35	25	15	—
1989	15.94	40	20	20	—

[a] in millions US$.

control of ventricular arrhythmias. Severe complications related to the ablation procedure suggest that this approach should be restricted to patients with very symptomatic and drug-refractory supraventricular tachyarrhythmias.

Since 1980, life-threatening tachycardias or fibrillation can be terminated by direct-current countershock provided by implantable cardioverter-defibrillators. The AICD was conceived by Mirowski as an implantable device for the detection and termination of ventricular fibrillation and, since receiving FDA approval, more than 10,000 implants of the device have been performed. The AICD demonstrates high accuracy in identifying both ventricular fibrillation and ventricular tachycardia and survival rates exceed those of patients treated with drug therapy for a comparable high-risk population.

Even if an analysis of the cost-effectiveness of the implantable defibrillator concludes that an approximate cost of $US17,100 per year-life saved is within an acceptable range compared to other life-saving interventions [7], some reserves have to be made as far as the explosive expansion of the implantation rate is concerned.

The effect on overall mortality is a lot more controversial than the widely recognized impact of the AICD on the subsequent risk of sudden death. Careful preoperative evaluation, EP testing, and patient selection remain decisive elements for the successful use of all antitachycardia devices [8].

A greater longevity of the device, combination with antitachycardia pacing and programmability, would not necessarily provide the AICD with better cost-effectiveness, because of the higher prices of newer devices and the side-effects of a longer duration of ventricular fibrillation with more postshock hemodynamic depression caused by longer episodes of arrhythmia before shocks are delivered [9].

Socio-economic cost-effectiveness is also irrelevant when reemployment following implantation of the AICD is considered. Only 29 patients out of 101 who received an AICD and, regarding their age, could have been professionally active prior to the implantation, returned to work after 11 ± 9 weeks followup. Moreover, a multivariate analysis indicated the level of education and marital status as best predictors of reemployment status, so that nonmedical factors play an important role in the resumption of work-related activities [10].

Perhaps there should be more invested and renewed interest in the organized use of automatic external defibrillators as an additional option in the treatment of out-of-hospital ventricular tachycardia in factories, offices, public buildings and at home. This can result in a significant increase of lives saved from sudden cardiac death.

In this country, there might be room for 100 carefully selected AICD implantations a year, which together with a growing replacement market represents a total cost of BF 150 million.

The limitation of AICD reimbursement to a restricted number of recognized treatment centres and on previously agreed well-defined conditions, as documented ventricular fibrillation or unstable ventricular tachycardia not responding to drug therapy and not indicated for interventional therapy, has proven to be an effective way of managing health insurance expenditure for the device in Belgium, where the implantation centres have accepted to perform a mutual peer review and a yearly reimbursed quotum of AICD's — of which every participating centre has its own share according to its needs — is fixed.

Extension of AICD implantation in Belgium with full insurance cover can be justified in the near future for survivors of sudden cardiac death when early implantation proves to be cost-effective compared with drug therapy guided by serial EP testing [11] and reuse of refurbished apparatus with maintained health care standards will be accepted for certain indications.

HEALTH INSURANCE POLICY

In Belgium, health-care delivery is financed by social security through the reimbursement of patient hospitalization days to the hospital and of diagnostic and therapeutic procedures through a fee-for-service system to the health professional. It is clear that this kind of output-financed health-care system can lead to an increase of services offered and a generalized purchase of the latest and most expensive technology [12].

The health-care policy is strongly influenced by pressure groups, but the fundamental reasons for state interference in a country with solidarity-based compulsory insurance for every citizen are obvious: careful observation of a maximum percentage of the GNP to be spent, guaranteed access rates to care for every member of the society and maintained quality care in setting out rules and conditions to the provision of health care.

To achieve these goals, health insurance relies mainly on three advisory bodies. The National Council of Hospitals advises the Minister of Health and Social Security on the planning, financing, or subsidizing of hospital infrastructure and facilities. The Medical Technical Council proposes the reimbursement of new medical technology based on a consensus between representatives of the medical schools, the providers, and the insurers. Special conditions of provision, fees and reimbursement rates are determined in this concertation body. Finally, the Technical Council for Implants and Devices can propose reimbursement for medical materials agreed for marketing by the Ministry of Health and based upon acceptance by carers, industry, and insurers.

Health insurers are not against appropriate medical technology, but intervention in the sacrosanct patient-provider relationship is essential in order to reimburse only effective technology, to promote quality, and to guarantee freedom of access.

CONCLUSION

The increase of the highest age groups within the overall greying population with more chronic and invalidating pathology, will certainly intensify the need for medical care together with the need for a greater application of new technologies.

The influences of manpower, hospital capacity, and technology on health-cost explosions are strongly interrelated regardless of the model of health-care financing. In order to provide further insurance cover for quality care and new medical technology, a well organized assessment capability is needed, because the insurer is not seeking a-priori limitation of development or diffusion of health technology.

In arrhythmia technology such as drugs, surgical intervention, or implantable devices, the health insurance has to render new technologies accessible at the appropriate time by adapting its reimbursement policy and so constantly inciting the substitution of less effective or obsolete diagnostic and therapeutic tools.

Evaluation of new procedures in the detection and treatment of arrhythmia for health insurance can never be based on the direct costs of a product or device used, but has to rely on the results of medical, biotechnological, and epidemiological research. All relevant costs related to invested and prevented health care or to direct and indirect expenses for the patient, his environment, and third parties, should be included. Quality assurance by peer review, reimbursed use of highly specialized and expensive techniques reserved to cardio-vascular centers with appropriate qualification and training and continued postgraduate education for the family physician to obtain correct and timely referrals, are complementary to the efforts to control health-care expenditures.

Manufacturers of drugs, cardiac pacemakers, and AICD must aim at ongoing improvement of their products and offer only to adequately trained doctors, cost-effective treatment facilities that can be accepted by both the insurer and the patient.

Improving arrhythmia technology means more specific action, fewer side effects, greater longevity, and multifactorial treatment, while acceptability has to do with a positive cost-utility ratio, possible alternatives, reuse, and substitution. The combination of these measures will permit the delivery and reimbursement of good medical care in the future.

REFERENCES

1. Goldschlager N (1988) Permanent cardiac pacing for brady-arrhythmia. *Postgrad Med* 83: 156—174.
2. Serwer GA, Mericle JM (1987) Evaluation of pacemaker pulse generator and patient longevity in patients aged 1 day to 20 years. *Am J Cardiol* 59: 824—827.
3. Parsonnet V, Berstein AD, Perry GY (1990) The nuclear pacemaker: is renewed interest warranted? *Am J Cardiol* 66: 837—842.
4. Kadish A, Calkins H, de-Buitleir M, Morady F (1990) Feasibility and cost savings of outpatient electrophysiologic testing. *J Am Coll Cardiol* 16: 1415—1419.
5. Lezaun R, Brugada P, Smeets J, et al. (1989). Cost-benefit analysis of medical vs surgical treatment of symptomatic patients with accessory atrioventricular pathways. *Eur Heart J* 10: 1105—1109.
6. Gallagher JJ, Selle JG, Sevenson RH, et al. (1988) Surgical treatment of arrhythmias. *Am J Cardiol* 61: 27A—44A.
7. Kuppermann M, Luce BR, McGovern B, Podrid PJ, Bigger JT, Ruskin JN (1990) An analysis of the cost effectiveness of the implantable defibrillator. *Circulation* 81: 91—100.
8. Kay GN, Plumb VJ, Dailey SM, Epstein AE (1990) Current role of the automatic implantable cardioverter-defibrillator in the treatment of life-threatening ventricular arrhythmias. *Am J Med* 88: 25H—34N.
9. Winkle RA, Mead RH, Ruder MA, Smith NA, Buch WS, Gaudiani VA (1990) Effect of duration of ventricular fibrillation of defibrillation efficacy in humans. *Circulation* 81: 1477—1481.
10. Kalbfleisch KR, Lehmann MH, Steinman RT, et al. (1989) Reemployment following implantation of the automatic cardioverter defibrillator. *Am J Cardiol* 64: 199—202.
11. O'Donoghue S, Platia EV, Brooks-Robinson S, Mispireta L (1990) Automatic implantable cardioverter-defibrillator: is early implantation cost-effective? *J Am Coll Cardiol* 16: 1258—1263.
12. van den Oever R (1986) The impact of health technology on funds, pp. 47—56 in *Proceedings of IFVHSF Conference, 1986.* London: International Federation of Voluntary Health Service Funds.

Index